BestMasters

Springer awards „BestMasters" to the best master's theses which have been completed at renowned universities in Germany, Austria, and Switzerland.

The studies received highest marks and were recommended for publication by supervisors. They address current issues from various fields of research in natural sciences, psychology, technology, and economics.

The series addresses practitioners as well as scientists and, in particular, offers guidance for early stage researchers.

Thomas Ondra

Optimized Response-Adaptive Clinical Trials

Sequential Treatment Allocation Based on Markov Decision Problems

 Springer Spektrum

Thomas Ondra
Vienna, Austria

BestMasters
ISBN 978-3-658-08343-4   ISBN 978-3-658-08344-1 (eBook)
DOI 10.1007/978-3-658-08344-1

Library of Congress Control Number: 2014957153

Springer Spektrum

Printed on acid-free paper

Springer Spektrum is a brand of Springer Fachmedien Wiesbaden
Springer Fachmedien Wiesbaden is part of Springer Science+Business Media
(www.springer.com)

Danksagung

An dieser Stelle möchte ich mich bei all jenen bedanken, die mich bei der Fertigstellung meiner Masterarbeit unterstützt haben.

Mein Dank gilt vor allem meinem Betreuer Prof. Hermann Schichl, welcher durch seine hervorragenden Vorlesungen mein Interesse an angewandter Mathematik geweckt hat. Prof. Hermann Schichl und Prof. Arnold Neumaier hatten, insbesondere im Rahmen ihres Seminares, jederzeit ein offenes Ohr, um etwaige Probleme zu besprechen. Weiters möchte ich mich einerseits für die spannende Themenstellung und andererseits für die Freiheit, diese zu bearbeiten, bedanken.

Mein besonderer Dank gilt auch meinen Eltern Andrea und Rudolf Ondra, die mir mein Studium erst ermöglicht haben und mich jederzeit in all meinen Entscheidungen unterstützt haben.

Besonders möchte ich Prof. Martin Posch, Leiter des Institutes für Medizinische Statistik an der Medizinischen Universität Wien, danken, welcher mich auf die Möglichkeit, klinische Studien im Blickwinkel der dynamischen Programmierung zu sehen, aufmerksam gemacht hat und das Entstehen dieser Arbeit durch sein FWF-Projekt P23167 *"Testen und Schätzen bei adaptiven Designs mit verblindeten und unverblindeten Interimanalysen"* unterstützt hat.

Thomas Ondra

Abstract

In this thesis we model a response-adaptive clinical trial as a Markov decision problem. Patients are treated sequentially and allocation of a study participant to a treatment is allowed to depend on the previous outcomes. To do so we present the main solution techniques for Markov decision problems with finite and infinite time horizon and give some examples. Then we discuss how a model for a clinical trial can be constructed. Our goal is to detect the superior treatment and keep the number of patients receiving the inferior treatment small. Finally, we carry out some simulations and compare the technique with an equal randomization strategy.

In the *first chapter* we give an introduction to Markov decision problems and consider a few examples to get familiar with some basic concepts.

The *second chapter* presents the basic solution theory for finite horizon Markov decision problems under the total reward criteria. We define the Bellman equations which are an optimality condition for Markov decision problems and have a look at the Backward Induction Algorithm. Finally, we solve the examples introduced at the beginning.

In the *third chapter* we take a look at the infinite horizon case. We work through the relationship between the solution of the Bellman equations for the infinite horizon case and an optimal policy. Then we consider the two most important procedures solving infinite horizon Markov decision problems: Value Iteration and Policy Iteration. We prove the convergence of both algorithms.

Chapters 1, 2 and 3 mainly follow the book [Put94].

The *fourth chapter* connects Markov decision problems with clinical trials and is based on [Pre09] and [BE95] where bandit models for clinical trials are introduced. We assume that the trial members are treated sequentially and the response is observed immediately. The main idea in these articles is that allocation to one of the treatments is based on what is learned so far so that

1. the superior treatment is detected in the end and

2. there are only few patient losses due to learning effects.

We investigate the applicability of the solution techniques discussed in the previous chapters and provide some numerical results. These techniques are particularly suitable for rare diseases since the patients are often in a live-threatening situation and clinicians do not have the possibility to rely on big clinical trials. Furthermore we compare the outcomes to an equal randomization strategy.

Zusammenfassung

In dieser Arbeit modellieren wir klinische Studien als Markov-Entscheidungsprobleme. Patienten werden sequentiell behandelt und die Zuordnung eines Patienten zu einem Medikament darf von den bereits erlernten Behandlungsergebnissen der vorherigen Studienteilnehmer abhängen. Dazu erarbeiten wir zunächst die grundlegenden Lösungsmethoden für Markov-Entscheidungsprobleme mit endlichem und unendlichem Zeithorizont und besprechen einige Beispiele. Dann beschreiben wir, wie ein Modell für eine klinische Studie konstruiert werden kann. Das Ziel ist dabei, das bessere Medikament zu erkennen und gleichzeitig die Anzahl der Studienteilnehmer, welche das schlechtere Medikament bekommen, gering zu halten. Schlussendlich führen wir einige numerische Simulationen durch und vergleichen die Ergebnisse mit einer Randomisierungsstrategie.

Im ersten Kapitel geben wir zunächst eine Einführung in Markov-Entscheidungsprobleme. Danach besprechen wir einige Beispiele, um mit den Grundkonzepten dieser Entscheidungsprobleme vertraut zu werden.

Im zweiten Kapitel erarbeiten wir die zugrunde liegende Lösungstheorie für Markov-Entscheidungsprobleme mit endlichem Zeithorizont unter dem Gesamtgewinn-Kriterium. Dazu definieren wir die Bellman-Gleichungen, welche als Optimalitätskriterium für Markov-Entscheidungsprobleme aufgefasst werden können, und beschreiben den Rückwärts-Induktionsalgorithmus. Schließlich lösen wir mit seiner Hilfe die anfangs beschriebenen Beispiele.

Im dritten Kapitel betrachten wir den Fall des unendlichen Zeithorizontes. Wir erarbeiten den Zusammenhang zwischen den Bellman–Gleichungen für den Fall des unendlichen Zeithorizontes und einer optimalen Entscheidungsregel. Danach beschreiben wir die zwei wichtigsten Lösungsverfahren: Die Werte-Iteration und die Entscheidungsregel-Iteration. Wir beweisen die Konvergenz beider Algorithmen.

Kapitel 1, 2 und 3 basieren in großen Teilen auf dem Buch [Put94]. Das vierte Kapitel verbindet Markov-Entscheidungsprobleme und klinische Studien und basiert auf den Arbeiten [Pre09] und [BE95], wo Banditen-Modelle für klinische Studien vorgestellt werden. Wir nehmen an, dass die Teilnehmer der Studie nacheinander behandelt werden und die Auswirkung der Behandlung unmittelbar danach messbar ist. Die Idee der oben genannten Arbeiten ist, dass die weitere Zuordnung von Patienten zu Behandlungen auf dem bereits Gelernten basiert und zwar so, dass

1. die bessere Behandlung gefunden wird, und

2. nur wenige Patienten, um einen Lerneffekt zu erzielen, die schlechtere Behandlung bekommen.

Wir untersuchen die Anwendbarkeit der in den vorigen Kapiteln vorgestellten Lösungsmethoden und führen einige numerische Simulationen durch. Die präsentierten Techniken sind vor allem für seltene Krankheiten interessant, da sich die Patienten oft in einer lebensbedrohlichen Situation befinden und Kliniker nicht auf große klinische Studien zurückgreifen können. Weiters vergleichen wir die Ergebnisse mit einer Randomisierungsstrategie.

Contents

List of Figures

List of Tables

1 Introduction to Markov Decision Problems

We take a look at the basic ingredients of Markov decision problems and introduce some basic notion. Then we construct the underlying stochastic model, which provides an appropriate framework for comparing the value of two policies. Finally, to get familiar with the matter, we give some examples of Markov decision problems: we analyse one period Markov decision problems, discuss a card game, and we explain how a single product stochastic inventory control problem can be modelled within the presented framework. This chapter is based on the book [Put94].

1.1 Basic Notions

Consider a controller who has the ability to influence a stochastic system by choosing actions at specified time points, called *decision epochs*. With T we denote the set of all decision epochs, here T is assumed to be finite or countably infinite. In the first case the decision problem is called a *finite horizon* problem, in the latter case its called an *infinite horizon* problem. At every decision epoch $t \in T$ the system attends a state $s \in S$, where S denotes the set of all possible states. Here S is assumed to be a countable set. Then the controller performs an action $a \in A_s$ where A_s is the set of all actions allowed when the system is in state s. With A we abbreviate the set of all possible actions, $A = \bigcup_{s \in S} A_s$. By choosing an action the controller can influence the behaviour of the stochastic system. Depending on the current state s and the action a performed at time t the controller gains a reward $r_t(s, a)$ and the system attends a new state σ, based on a probability distribution $p_t(\cdot|s,a)$ which is called the *transition probability* function. So $p_t(\sigma|s,a)$ is the probability that the next state equals $\sigma \in S$ given the controller chooses action a and the state at time t equals s.

Often the reward does not only depend on the current state s and the action a but also on the next state σ. We do not know σ in advance since it is random, so we calculate the expected reward

$$r_t(s,a) := \sum_{\sigma \in S} p_t(\sigma|s,a) r_t(s,a,\sigma),$$

where $r_t(s,a,\sigma)$ is the reward gained when the systems current state is s, the controller chooses action a and $\sigma \in S$ is the unknown next state.

When we deal with a finite horizon problem we do not have to choose an action at time point N, so the final reward $r_N(s)$ is only dependent on the last state. Now we can define a *Markov decision process*.

Definition 1.1 (Markov decision process). *The quintuple*

$$(T, S, A_s, p_t(\cdot|s,a), r_t(s,a))$$

defines a Markov decision process. *Here T is the set of decision epochs, S is the set of all possible states, A_s is the set of actions allowed when the system is in state s, $p_t(\cdot|s,a)$ are transition probabilities and $r_t(s,a)$ are the expected reward functions for all $t \in T$.*

Remark 1.2. We use the term "Markov" since the reward functions and the transition probabilities only depend on the past through the current state and the current action.

A *decision rule* specifies the choice of an action. In the easiest case we deal with a *deterministic Markovian* decision rule, this is a function $d_t : S \to A_s$ which determines the action to be taken at time point t when the system is in state s.

Then there are *deterministic history dependent* decision rules. A deterministic history dependent decision rule is a function depending not only on the current state but also on all states occurred, and on all actions chosen so far. We define $s_{1:t}$ as the vector of all states until decision epoch t. With $a_{1:t-1}$ we denote the vector of all actions taken until decision epoch t. Now we call $h_t := (s_{1:t}, a_{1:t-1})$ the history until t. In the deterministic history dependent case the action a_t is specified by a function $d_t : H_t \to A_s$ where H_t is the set of all possible histories and s is the state in decision epoch t.

Remark 1.3. Sometimes it is convenient to rearrange states and actions in the history h_t according to $h_t = (s_1, a_1, s_2, \ldots, a_{N-1}, s_N)$. Then H_t satisfies the recursion $H_{t+1} = H_t \times S \times A$. Especially in Chapter 2 we will use this notation.

In addition to the deterministic case we deal with *randomized* decision rules. With $\mathcal{P}(A_s)$ we denote the set of all probability measures on A_s. A *randomized Markovian* decision rule selects an element $q(\cdot) \in \mathcal{P}(A_s)$, so a randomized Markovian decision rule is an operator $d_t : S \to \mathcal{P}(A_s)$ where s is the state of the system at decision epoch t.

Then we have *randomized history dependent* decision rules. This is an operator $d_t : H_t \to \mathcal{P}(A_s)$ where s is again the current state of the system.

Remark 1.4. If we want to stress that we work with a randomized Markovian decision we use $q_{d_t(s)}$ with $s \in S$, whereas we use $q_{d_t(h)}$ with $h \in H_t$ to emphasize that we are working with a randomized history dependent decision rule. Whenever this is not necessary we simply write q_t.

Randomized decision rules are a generalization of deterministic decision rules since we can identify a deterministic decision rule with a degenerate probability distribution, that is $q_{d_t(s)}(a) = 1$ for an $a \in A_s$ in the Markovian and $q_{d_t(h)}(a) = 1$ for an $a \in A_s$ in the history dependent case.

We define the set of decision rules available at time t with D_t^K, where K stands for one of the discussed decision rules,

$$K \in \{DM, DH, RM, RH\}.$$

Here DM stands for deterministic Markovian, DH for deterministic history dependent, RM for randomized Markovian and RH for randomized history depended.

By choosing $d_t \in D_t^{DM}$ transition probabilities and reward functions become functions only depending on the states, $p_t(\cdot|s, d_t(s))$ and $r_t(s, d_t(s))$. If we choose $d_t \in D^{RM}$ we calculate the transition probabilities and the rewards by

$$p_t(\sigma|s, d_t(s)) = \sum_{a \in A} p_t(\sigma|s, a) q_{d_t(s)}(a),$$

$$r_t(s, d_t(s)) = \sum_{a \in A} r(s, a) q_{d_t(s)}(a).$$

Hence we simply weight the transition probabilities and the rewards with the selected probability measure $q_{d_t(s)}$.

Choosing a history dependent decision rule means that the transition probabilities and the rewards depend on the history H_t. If $d_t \in D_t^{DH}$ we denote the rewards by $r_t(s, d_t(h))$ and the transition probabilities by $p_t(\cdot|s, d_t(h))$. If $d_t \in D_t^{RH}$ transition probabilities and rewards are calculated by

$$p_t(\sigma|s, d_t(s)) = \sum_{a \in A} p_t(\sigma|s, a) q_{d_t(h)}(a), \tag{1.1a}$$

$$r_t(s, d_t(h)) = \sum_{a \in A} r(s, a) q_{d_t(h)}(a). \tag{1.1b}$$

A *policy* π is a sequence of decision rules, $\pi = (d_t)_{t=1}^{N-1}$ in the finite horizon case and $\pi = (d_t)_{t=1}^{\infty}$ in the infinite horizon case. The index t only reaches from one to $N - 1$ in the finite horizon case since by definition we do not choose an action in the last decision epoch N.

A policy is a complete description for the controller what to do in every single decision epoch $t \in T$. With $\mathcal{D}^K$ we denote the set of all possible policies, $\mathcal{D}^K := \prod_{t=1}^{\infty} D_t^K$ in the infinite horizon case and $\mathcal{D}^K := \prod_{t=1}^{N-1} D_t^K$ in the finite horizon case . Here K stands again for any kind of decision rule, $K \in \{DM, DH, RM, RH\}$. We have $\mathcal{D}^{DM} \subset \mathcal{D}^{DH} \subset \mathcal{D}^{RH}$ and $\mathcal{D}^{DM} \subset \mathcal{D}^{RM} \subset \mathcal{D}^{RH}$. So $\mathcal{D}^{RH}$ contains

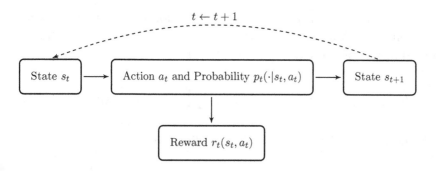

Figure 1.1: Overview of the components of a Markov decision process: Given the current state $s = s_t$ the controller chooses an action $a = a_t$ and receives an (expected) reward of $r_t(s_t, a_t)$. Then the state changes according to $p_t(\cdot | s_t, a_t)$ to the succeeding state $\sigma = s_{t+1}$.

the most general and $\mathcal{D}^{DM}$ contains the most specific policies. In Figure 1.1 we see an overview of the components of a Markov decision process.

1.2 Probabilities and Induced Stochastic Processes

Now we construct a stochastic model for Markov decision problems. Lets first assume a finite horizon. Lets define $\Omega_N := H_N = S^N \times A^{N-1}$. A typical $\omega \in \Omega$ now looks like

$$\omega = (s_{1:N}, a_{1:N-1}),$$

that carries the information which states have been attained and which actions have been played. Specifying a σ-algebra $\mathcal{A}(\Omega)$ is easy: we simply take the power set of Ω. Let now $P_1(\cdot)$ be the initial probability distribution of the first state. In applications a first state is often predetermined, in this case we can think of $P_1(\cdot)$ as a degenerated probability distribution.

Now let $\pi = (d_t)_{t=1}^{N-1} \in \mathcal{D}^{RH}$ be a randomized history dependent policy. A Markov decision process and the policy[1] π define a probability measure P_N^π on $(\Omega, \mathcal{A}(\Omega))$ through

$$P_N^\pi(\omega) = P_N^\pi(s_{1:N}, a_{1:N-1}) = P_1(s_1) \prod_{t=1}^{N-1} q_t(a_t) p_t(s_{t+1}|s_t, a_t). \quad (1.2)$$

For any deterministic policy (1.2) simplifies to

$$P_N^\pi(\omega) =$$
$$\begin{cases} P_1(s_1) \prod_{t=1}^{N-1} p_t(s_{t+1}|s_t, a_t) & \text{if } d_t(s_t) = a_t \; \forall t = 1, \dots, N-1 \\ 0 & \text{otherwise.} \end{cases}$$
$$(1.3)$$

Remark 1.5. Indeed P_N^π in (1.2) defines a probability measure: Consider two discrete probability measures (Ω_1, P_1) and (Ω_2, P_2). Then we can define a probability measure P on $\Omega_1 \times \Omega_2$ where $P(\omega) = P(\omega_1, \omega_2) := P_1(\omega_1) P_2(\omega_2)$. We have $0 \le P(\omega) \le 1$ since $0 \le P_1(\omega) \le 1$ and $0 \le P_2(\omega) \le 1$. Furthermore we see

$$\sum_{(\omega_1, \omega_2) \in \Omega_1 \times \Omega_2} P(\omega) = \sum_{\omega_1 \in \Omega_1} \sum_{\omega_2 \in \Omega_2} P_1(\omega_1) P_2(\omega_2)$$
$$= \sum_{\omega_1 \in \Omega_1} P_1(\omega_1) \sum_{\omega_2 \in \Omega_2} P_2(\omega_2)$$
$$= 1.$$

Since $p_t(\cdot|s_t, a_t)$, $q_t(\cdot)$, and $P_1(\cdot)$ are by definition probability measures consecutively applying this fact proves the claim.

Now we define the random variables $X_t : \Omega \to S$, $Y_t : \Omega \to A$ and $Z_t : \Omega \to H_t$ by $X_t(\omega) := s_t$, $Y_t(\omega) := a_t$ and $Z_t(\omega) := h_t = (s_{1:t}, a_{1:t-1})$. We have

$$P_N^\pi(X_1 = s) = P_1(s) \qquad (1.4a)$$

---

[1]Since randomized history dependent policies are the most general ones any policy discussed previously clearly defines such a probability measure.

$$P_N^\pi(Y_t = a | Z_t = h_t) = q_{d_t(h_t)}(a) \qquad (1.4\text{b})$$

$$P_N^\pi(X_{t+1} = \sigma | Z_t = h_t, Y_t = a_t) = p_t(\sigma | s_t, a_t) \qquad (1.4\text{c})$$

For example the first equality can be verified as follows:

$$
\begin{aligned}
P_N^\pi(X_1 = s) &= P(\{(s_1, a_1, \ldots, s_N) \in \Omega : s_1 = s\}) \\
&= \sum_{a_1, \ldots, s_N} P(\{(s, a_1, \ldots, s_N)\}) \\
&= P_1(s) \sum_{a_1, \ldots, s_N} \prod_{t=1}^{N-1} q_t(a_t) p_t(s_{t+1} | s_t, a_t) \\
&= P_1(s).
\end{aligned}
$$

The last equality follows because the product inside the sum defines a probability measure on $\tilde{\Omega} := \{(a_1, s_2, \ldots, s_N) : a_i \in A, s_i \in S\}$. We sum the probabilities over the entire space $\tilde{\Omega}$ which has to be equal to one and hence (1.4a) follows. Equations (1.4b) and (1.4c) follow in a similar way.

Definition 1.6 (Markov chain). *Let (Ω, P) be a discrete probability space and $X_t : \Omega \to S$ with $t \in \mathbb{N}$ a family of random variables where S is any countable set. We say that $(X_t)_{t \in \mathbb{N}}$ is a Markov chain if*

$$
\begin{aligned}
P(X_{t+1} = s | X_t = s_t, X_{t-1} &= s_{t-1}, \ldots, X_1 = s_1) \\
&= P(X_{t+1} = s | X_t = s_t)
\end{aligned} \qquad (1.5)
$$

for all $t \in \mathbb{N}$.

Now for any Markovian decision rule π, deterministic or history dependent, the family X_t defines a Markov chain on $(\Omega, \mathcal{A}, P_N^\pi)$. As we see in (1.4c) the next state is only dependent on the current state and the current action played. The current action itself in the Markovian policy case only depends on the current state because equation (1.4b) simplifies in that case to $P_N^\pi(Y_t = a | Z_t = h_t) = q_{d_t(s_t)}(a)$. This justifies the name Markov decision process. For history dependent policies the action depends on the history as we see in (1.4b). This destroys the *Markov chain* feature.

Now let W be an arbitrary real valued random variable on the probability space $(\Omega, \mathcal{A}, P_N^\pi)$. The expected value of W is defined by

$$E^\pi(W) := \sum_{\omega \in \Omega} P_N^\pi(\omega) W(\omega) = \sum_{w \in \mathbb{R}} w P_N^\pi(\omega : W(\omega) = w).$$

Often we will have $W = \sum_{t=1}^{N-1} r_t(X_t, Y_t) + r_N(X_N)$, the sum of the expected rewards at every decision epoch and the expected final reward.

1.2.1 Conditional Probabilities and Conditional Expectations

Let $h_t = (\overline{s_1}, \overline{a_1}, \dots, \overline{s_t})$ be a history up to time t. We define the set of all sample paths with the history h_t, $\Omega(h_t) := \{(s_1, a_1, \dots, s_N) \in \Omega : s_1 = \overline{s_1}, \dots, s_t = \overline{s_t}\}$. Furthermore, we define the conditional probability (conditioned on the history h_t) by

$$P_N^\pi(a_t, s_{t+1}, \dots, s_N | h_t) := \begin{cases} \frac{P_N^\pi(s_1, a_1, \dots, s_N)}{P_N^\pi(\Omega(h_t))} & \text{if } P_N^\pi(\Omega(h_t)) \neq 0 \\ 0 & \text{otherwise.} \end{cases}$$

Finally we define the expectation conditioned on the history h_t of a random variable $W : \Omega \to \mathbb{R}$ through

$$E_{h_t}^\pi(W) := \sum_{\omega \in \Omega(h_t)} P_N^\pi(a_t, s_{t+1}, \dots, s_N | h_t) W(\omega).$$

Now we have a suitable framework for the finite horizon case.

Lets consider the infinite horizon case. We define the sample space by

$$\Omega := \{\omega = (s_1, a_1, s_2, a_2, s_3, \dots) : s_j \in S, a_j \in A_{s_j}\}.$$

In such a situation there exists a standard construction for a suitable σ-algebra:

Definition 1.7. *Let Ω be a set and let M be any subset of the power set of Ω. We define*

$$\sigma(M) := \bigcap_{\mathcal{A}: M \subseteq \mathcal{A}} \mathcal{A}$$

as the intersection of all σ-algebras $\mathcal{A}$ over Ω containing M.

It is easy to verify that $\sigma(M)$ is indeed a σ-Algebra over Ω. Now consider the event

$$F^N_{s_{1:N},a_{1:N-1}} :=$$
$$\{\omega \in \Omega : \omega_1 = s_1, \omega_2 = a_1, \omega_3 = s_1, \ldots, \omega_{2N} = a_{n-1}, \omega_{2N-1} = s_N\}$$

and define

$$M = \bigcup_{N \in \mathbb{N}} \bigcup_{s_{1:N} \in S^N} \bigcup_{a_{1:N-1} \in A_{s_1} \times \cdots \times A_{s_{N-1}}} F^N_{s_{1:N},a_{1:N-1}}.$$

Now we use the σ-algebra $\mathcal{A} := \sigma(M)$. Again any history dependent randomized policy π defines a probability measure $P^\pi : \mathcal{A} \to \Omega$ through

$$P^\pi(F^N_{s_{1:N},a_{1:N-1}}) := P^\pi_N(s_1, a_1, \ldots, s_N),$$

where the right hand side is defined in (1.2).

1.3 Examples

We will start with analysing one period Markov decision problems to get familiar with some basic concepts. Then we will have a look at examples to see how to construct finite Markov decision problems.

One Period Markov Decision Problems

Let A and S be finite sets and $T = \{1, 2\}$. Here we have $\Omega = S \times A \times S$ and $\Omega \ni \omega = (s_1, a_1, s_2)$. We assume that the system is in a predetermined initial state s_1. Then we have to choose an action $a_1 \in A$ to receive an immediate reward of $r_1(s_1, a_1)$. Then the system changes to state s_2 depending on $p_1(\cdot|s_1, a_1)$. In a one-period Markov decision process this is already our final state and we receive a final reward of $r_2(s_2)$. Our goal is now to choose the best possible action a_1. Therefore, we have to find a suitable optimality criterion which reflects our idea of a_1 being a good choice. A Markov decision process together with such an optimality criterion is called a *Markov decision problem*. We want to find an a_1 which maximizes the sum of the immediate reward $r_1(s_1, a_2)$ and the expected final reward $r_2(s_2)$. Therefore any

$$a^* \in \underset{a \in A_{s_1}}{\text{argmax}} \ r_1(s_1, a) + \sum_{\sigma \in S} p_1(\sigma|s_1, a) r_2(\sigma) \tag{1.6}$$

is optimal. If we want to find an optimal policy $\pi \in \mathcal{D}^{DM}$ we have to do the maximization step in (1.6) for every possible starting state $s_1 \in S$ to find the optimal decision rule $d_1 : S \to A$. Then, since it is only a one-period Markov decision process, we already found the optimal policy $\pi = d_1$. We cannot illustrate history dependent decision rules because the history until our first and only decision a_1 is $h_1 = s_1$, therefore there is nothing new.

Randomized policies $\pi \in \mathcal{D}^{RM}$ do not lead to a better reward. Here we have to solve

$$\max_{q \in \mathcal{P}(A_{s_1})} \sum_{a_1 \in A} q(a_1) \left(r_1(s_1, a_1) + \sum_{\sigma \in S} p(\sigma|s_1, a_1) r_2(\sigma) \right),$$

where we maximize over all probability measures on A_{s_1}. Since $0 \le q(\cdot) \le 1$ we have

$$\max_{q \in \mathcal{P}(A_{s_1})} \sum_{a_1 \in A} q_1(a_1) \left(r_1(s_1, a_1) + \sum_{\sigma \in S} p(\sigma|s_1, a_1) r_2(\sigma) \right)$$
$$= \max_{a_1 \in A} r_1(s_1, a_1) + \sum_{\sigma \in S} p(\sigma|s_1, a_1) r_2(\sigma). \tag{1.7}$$

So we cannot do better with randomized strategies. Of course a^* in (1.6) does not need to be unique. Any randomized strategy which sets a positive weight only to actions fulfilling (1.6) are optimal as well.

The (k, n) Card Game

Let us consider the following easy Black Jack like card game. Altogether we have k cards, numbered from 1 to k. The cards are upside down and the player has to randomly choose one card. The number $x_1 \in \{1, \ldots, k\}$ of the first card is written down. Then the chosen card is *laid back* and the player has to choose a second card. The number x_2 of the new card is added to the first card number, $s = x_1 + x_2$. This procedure is iterated and the card values are summed up. The goal is to play in such a way that the total sum s is smaller than and as close as possible to a predefined number $n > k$. After the player decides to stop he receives a reward

$$
R = \begin{cases} s & \text{if } s \le n \\ 0 & \text{otherwise.} \end{cases}
$$

We want to solve the game using a suitable finite Markov decision problem. We take $N = n$, because the smallest card number is one and after n rounds the player definitely does not want to play on because either he was lucky and $s = n$ or, more likely, $s > n$. In the first case he receives the maximal reward $R = n$ in the second case he gains nothing because then we have $R = 0$. Clearly we only have two possible actions in every decision epoch, namely a_p which stands for "take another card" and a_n which stands for "take no further card". We define the current state s as the sum of the already chosen card values. This leads to the set S of all possible states, $S = \{1, \ldots, kn\}$. Taking no card means that $s' = s$ since the sum does not change. Taking a further card means that the state s' will be between $s + 1$ and $s + k$. We assume a predetermined initial state. This is the number of the first card drawn. Then we choose the following transition probabilities

$$
p_t(\sigma|s, a_p) = \begin{cases} \frac{1}{k} & \text{for } \sigma = s + j, \ 1 \le j \le k \\ 0 & \text{otherwise,} \end{cases}
$$

and

$$
p_t(\sigma|s, a_n) = \begin{cases} 1 & \text{for } \sigma = s \\ 0 & \text{otherwise.} \end{cases}
$$

Example 1.8. A short example illustrates how we define the probabilities. Take $n = 4$ and $k = 2$. Then we can write down the values of

$p_t(\sigma|s, a_p)$ in a matrix $P \in \mathbb{R}^{|S| \times |S|}$, since the action is fixed and the probabilities are time independent,

$$P = \begin{pmatrix} 0 & 0 & 0 & 0 & 0 & 0 & 0 & 0 \\ 0.5 & 0 & 0 & 0 & 0 & 0 & 0 & 0 \\ 0.5 & 0.5 & 0 & 0 & 0 & 0 & 0 & 0 \\ 0 & 0.5 & 0.5 & 0 & 0 & 0 & 0 & 0 \\ 0 & 0 & 0.5 & 0.5 & 0 & 0 & 0 & 0 \\ 0 & 0 & 0 & 0.5 & 0.5 & 0 & 0 & 0 \\ 0 & 0 & 0 & 0 & 0.5 & 0.5 & 0 & 0 \\ 0 & 0 & 0 & 0 & 0 & 0.5 & 0.5 & 0 \end{pmatrix}.$$

Here we have $P_{ij} := p_t(i|j, a_p)$, so P_{ij} is the probability that the next state is i if the current state is j and action a_p (take another card) is chosen.

Example 1.9. In Figure 1.2 we see all possible paths of the $(2, 4)$ card game with starting state $s_1 = 1$. We distinguish between grey and black arcs. Grey arcs are only drawn because the underlying MDP in principal does not a priori exclude these paths although a clever player would not choose them. Note that if $s_2 = 3$ then of course if the action a_p is chosen the state $s_3 = 3$ is not possible. Nevertheless, we have an arc between a_p and the state $s_3 = 3$ for the case that the state at time point two was equal to $s_2 = 2$. Knowing the whole past is not necessary for determining which states are possible at the next time point. The only two things which are necessary to know is the current action and the current state.

Now we have to specify the rewards. The player does not really get a reward in the decision epochs $t = 1, \ldots, N - 1$ so we choose $r_t(s, a_p) = r_t(s, a_n) = 0$. For the final rewards we choose $r_N(s) = s$ if $s \leq n$ and $r_N(s) = 0$ otherwise.

The most aggressive strategy would be to try to reach the maximal reward $R = n$ even if $s = n - 1$ after some cards have been drawn. The most conservative strategy would be to stop playing when $s > n - k$ for the first time. Somewhere in between there will be the strategy which yields the best possible expected reward. Since the intermediate

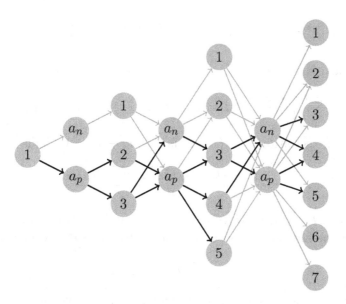

Figure 1.2: Overview of the $(2, 4)$ card game with starting state $s_1 = 1$.

rewards are all zero and we only have two possible actions we are also able to solve the problem without using Markov decision problem solving techniques. The fundamental question now is: Assume we are in state s, should we take another card or should we stop playing? To answer this question we define E_s as the expected reward if the current sum equals s and we decide to take another card. The second possibility would be not to choose a card and be satisfied with the assured reward s. Therefore it seems natural to solve

$$E_s \geq s.$$

Whenever this equation is fulfilled we better play on since we can expect a better reward than the assured reward s. Assume that the player already took cards such that $s > n - k$ is fulfilled for the fist time. We calculate

$$E_s = \frac{1}{k}((s + 1) + \ldots + n) = \frac{1}{k}\left(\frac{n(n + 1) - s(s + 1)}{2}\right)$$

$$= \frac{n^2 + n - s^2 - s}{2k}.$$

So we have

$$\frac{n^2 + n - s^2 - s}{2k} \geq s \Leftrightarrow s^2 + s(1 + 2k) + (-n^2 - n) \leq 0.$$

Solving the equation $s^2 + (1 + 2k)s + (-n^2 - n) = 0$ yields

$$s_{1,2} = \frac{-(1 + 2k) \pm \sqrt{(1 + 2k)^2 + 4(n^2 + n)}}{2}.$$

Because $n, k > 0$ we always have two distinct solutions. Let $s_1 > s_2$. Solving the inequality above now means to find $s \in \mathbb{R}$ which satisfy

$$(s \leq s_1 \wedge s \geq s_2).$$

Since we have $s_2 < 0$ and we consider sums of positive integers. So we get our solution set to $L = \{s \in \mathbb{R} : 0 \leq s \leq s_1 \}$. For these sums s we have an expected advantage do good if we decide to take a further card.

We now concretely use $n = 20, k = 10$ and assume that $s = 12$. Do we want to play on? Let us take a look at the expected reward if we decide to take another card, $E_{12} = \frac{1}{10}(20 + 19 + \ldots + 13) = 13.2$, so we should take another card since $13.2 > 12$. The situation changes if $s = 13$. We have $E_{13} = \frac{1}{10}(20 + 19 + \ldots + 14) = 11.9 < 13$ and therefore we're doing better in deciding not to take a further card. Indeed solving the equation $E_s \geq s$ with $n = 20, k = 10$ yields the solution set $L = \{s \in \mathbb{R} : 0 \leq s \leq 12.53\}$. This means if $s \leq 12$ we should play on and if $s \geq 13$ we should stop playing.

A short simulation affirms the results above. Every possible strategy is tested at 10000 simulated games. Under a possible strategy we understand a number α were we stop taking further cards if the sum s fulfils $s \geq \alpha$. There are more policies than possible strategies, but in the simulation we do not consider policies which stop playing and suddenly start taking cards again. There is no reason why a policy should not take the card right away. Looking at Figure 1.3 reveals that

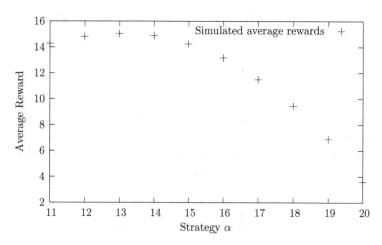

Figure 1.3: Simulated average rewards of the $(10, 20)$ card game if strategy α is used.

the simulation suggests the strategy $\alpha = 13$ since then the average outcome is maximal.

In the second chapter we will describe an algorithm for general finite horizon Markov decision problems. We will see that the solution suggested by MDP will also be $\alpha = 13$.

Stochastic Inventory Control

Consider a warehouse of capacity M which only stores one product type. Each month we are confronted with the task whether we want to stock additional units. If we decide to buy we also want to know how much we should purchase. Buying additional ware also means that some storage costs arise. Clearly we choose the state s to be the units currently stored in the warehouse. This leads to the state space $S := \{0, \ldots, M\}$. We define the set of all possible actions $A_s := \{0, \ldots M - s\}$, where an action $a \in A_s$ tells us how many units should be bought in addition. With D we denote the random demand until we again have the chance to reorder some units. This leads to the system equation

$$s' = \max\{0, s' + a - D\}.$$

We do not allow backlogging. It is assumed that if the demand exceeds our capacities the customer will buy as much as possible from our warehouse. We further assume that costs and rewards do not vary from decision epoch to decision epoch. Now we take a closer look at some economic parameters. Let the cost of ordering u units be

$$b(u) := \begin{cases} K + c(u) & \text{if } u > 0 \\ 0 & \text{otherwise,} \end{cases}$$

where $c(u)$ is, e.g., a linear function in u. The cost of maintaining u units is described by a non decreasing function $h(u)$. Finally, we receive a reward of $f(j)$ if we have been able to sell j units, where we assume that $f(0) = 0$. So altogether we have

$$r_t(s', s, a) = f(s + a - s') - b(a) - h(s + a).$$

Now we have to construct the expected rewards $r_t(s, a)$. Defining $p_j := P(D = j)$ we have that if our inventory exceeds the demand j we gain $f(j)$ with a probability of p_j at the end of the month. If the demand exceeds the inventory and we currently have u units stored we receive a reward of $f(u)$ with probability $q_u := \sum_{i \geq u} p_i$. This leads to the expected gain of

$$F(u) = q_u f(u) + \sum_{j=1}^{u-1} p_j f(j).$$

Now we are able to define the expected rewards as

$$r_t(s, a) := F(s + a) - b(a) - h(s + a)$$

for $t = 1, \ldots, N - 1$ and $r_N(s) = g(s)$ for some function $g(\cdot)$ which describes the value of the rest of the inventory in the last decision epoch. For example we could choose $g(s) = 0 \; \forall s \in S$ if for some reason after the last decision epoch it is impossible to sell further products. This might be the case when we exactly know that from some time point on there will be no further demand. Now we have to define the transition probabilities,

$$p_t(\sigma|s, a) := \begin{cases} 0 & M \geq \sigma > s + a \\ p_{s+a-\sigma} & M \geq s + a \geq \sigma > 0 \\ q_{s+a} & M \geq s + a \text{ and } \sigma = 0. \end{cases}$$

A short explanation follows. Note that $p_t(\sigma|s, a)$ describes the probability that we have σ units stored at time point $t + 1$ if we currently have s units stored and purchase a more units. Now we have to distinguish some cases. If $M \geq \sigma > s + a$ then $p_t(\sigma|s, a) = 0$ since it is not possible that the stock in the next period is greater than the sum of the currently stored units and the additionally bought units. We have $p_t(\sigma|s, a) = p_{s+a-\sigma}$ if $M \geq s + a \geq \sigma > 0$ since in the next epoch we want to have σ units left which means that we have to sell $s + a - \sigma$ units. The probability therefore equals $p_{s+a-\sigma}$. Finally, we have the case that the demand exceeds the inventory. The probability that at least $s + a$ units are sold equals q_{s+a}.

What about the time horizon? For example we could specify that we reorder at the beginning of every month and plan over a time horizon of ten months. This would give $N = 10$.

We defined the set of possible actions, the transition probabilities, the rewards and the time horizon and hence have a complete Markov decision problem formulation for a stochastic inventory control problem. At the end of Chapter 2 we will have a look at the optimal policy, which describes the optimal reordering strategy.

2 Finite Horizon Markov Decision Problems

In this chapter we solve finite horizon Markov decision problems. We are describing a policy evaluation algorithm and the Bellman equations, which are necessary and sufficient optimality conditions for Markov decision problems. Then we are constructing optimal policies out of the solution of the Bellman equations. We will see that the class of Markov deterministic policies —that are easier to handle—contain, under assumptions which are often satisfied in practise, optimal policies. Finally, we describe how optimal policies can be calculated, based on a backward induction algorithm. This chapter is based on [Put94], [Whi93], and [Der70].

2.1 Optimal Policies and the Bellman Equations

In order to be able to speak about optimal policies we need a method for comparing two policies. In the finite horizon case we can simply choose the so called expected reward criterion. Let π be a history dependent randomized policy and define

$$v_N^\pi(s) := E_s^\pi \left(\sum_{t=1}^{N-1} r_t(X_t, Y_t) + r_N(X_N) \right),$$

where the right hand side is the conditional expectation of the sum of the expected rewards $r_t(X_t, Y_t)$ and the final reward $r_N(X_N)$ conditioned on $X_1 = s$. Since the last reward $r_N(X_N)$ only depends on the last state and not on an action anymore, we have to write it down separately. Recall that we defined $X_t := s_t$ and $Y_t := a_t$. To be able to speak about expectations we need a suitable sample space $(\Omega, \mathcal{A}, P_N^\pi)$, which we defined for discrete A and S in Section 1.2. If we have $\pi \in \mathcal{D}^{DM}$ the action chosen at decision epoch t is determined by $d_t(s_t)$, so in this case we have

$$v_N^\pi(s) = E_s^\pi \left(\sum_{t=1}^{N-1} r_t(X_t, d_t(X_t)) + r_N(X_N) \right).$$

Definition 2.1 (Optimal policies and ε–optimal policies). *A policy π^* is called optimal if for every starting state $s \in S$ and all $\pi \in \mathcal{D}^{RH}$*

$$v_N^{\pi^*}(s) \geq v_N^\pi(s).$$

Fix $\varepsilon > 0$. We say that the policy π_ε^* is ε–optimal if for every starting state $s \in S$ and all $\pi \in \mathcal{D}^{RH}$

$$v_N^{\pi_\varepsilon^*}(s) + \varepsilon > v_N^\pi(s).$$

Definition 2.2 (Value of a Markov decision problem). *We define the value of a Markov decision problem as*

$$v_N^*(s) := \sup_{\pi \in \mathcal{D}^{RH}} v_N^\pi(s).$$

If S and A are finite and the rewards are bounded then the supremum exists and can be replaced by the maximum of the right hand side of the equation above. Clearly, in this case we have $v_N^*(s) = v_N^{\pi^*}(s)$, so the expected total reward equals the value of a Markov decision problem if an optimal policy π^* is used. If we use a ε–optimal policy we have $v_N^{\pi_\varepsilon^*}(s) + \varepsilon > v_N^*(s)$.

2.1.1 Policy Evaluation

Now we want to find a method which allows us to calculate the expected total reward $v_N^\pi(s)$ for a given policy π. We want to do this in a backward inductive way. We define functions $u_t^\pi : H_t \to \mathbb{R}$,

$$u_t^\pi(h_t) := E_{h_t}^\pi \left(r_N(X_N) + \sum_{n=t}^{N-1} r_n(X_n, Y_n) \right), \qquad (2.1)$$

the expected total reward from decision epoch t on given the history h_t up to time t. Furthermore we define $u_N^\pi(h_N) := r_N(s_N)$ where $h_N = (h_{N-1}, a_{N-1}, s_N)$. Note that if $h_1 = s$ and $t = 1$ we have

$$u_1^\pi(s) = E_s^\pi \left(r_N(X_N) + \sum_{n=1}^{N-1} r_n(X_n, Y_n) \right) = v_N^\pi(s). \qquad (2.2)$$

So if we are able to calculate $u_1^\pi(s)$ we know the value of the Markov decision problem if an optimal policy is used. Algorithm 1 calculates the functions u_t^π in a backward inductive way.

Algorithm 1 Finite Horizon Policy Evaluation Algorithm for $\pi \in \mathcal{D}^{RH}$

1: Set $u_N^\pi(h_N) = r_N(s_N)$ for each $h_N = (h_{N-1}, a_{N-1}, s_N) \in H_N$.
2: $t \leftarrow N$
3: **for** $t \neq 1$ **do**
4:    $t \leftarrow t - 1$
5:    Compute

$$u_t^\pi(h_t) =$$
$$\sum_{a \in A_{s_t}} q_{d_t(h_t)}(a) \left(r_t(s_t, a) + \sum_{\sigma \in S} p_t(\sigma | s_t, a) u_{t+1}^\pi((h_t, a, \sigma)) \right)$$

   for each $h_t \in H_t$.
6: **end for**

Of course we need to show that the $u_t^\pi(h_t)$ constructed by the algorithm are the same as these defined in (2.1).

Theorem 2.3. *Let π be a randomized history dependent policy and $u_t^\pi(h_t)$ be constructed by Algorithm 1. Then these functions are equal to the right hand side of (2.1) for all $t \leq N$, particularly $u_1^\pi(s)$ is the value of the underlying Markov decision problem if an optimal policy is used.*

Proof. We prove the claim with backward induction. Within this proof u_t^π always denotes the functions generated by Algorithm 1. We see that $u_N^\pi(h_N) = r_N(s_N)$ coincides with the definition in (2.1). Assume that $u_t^\pi(h_t) = E_{h_t}^\pi \left(r_N(X_N) + \sum_{k=t}^{N-1} r_k(X_k, Y_k) \right)$ is true for all $t = n+1, \ldots, N$. Now we calculate for $t = n$

$$
\begin{aligned}
u_n^\pi(h_n) =\ & \sum_{a \in A_{s_n}} q_{d_n(h_n)}(a) r_n(s_n, a) \\
& + \sum_{\sigma \in S} \sum_{a \in A_{s_n}} q_{d_n(h_n)} p_n(\sigma | s_n, a) u_{n+1}^\pi(h_n, a, \sigma) \\
\overset{(1.1)}{=}\ & r_n(s_n, d_n(h_n)) \\
& + \sum_{\sigma \in S} p_n(\sigma | s_n, d_n(h_n)) u_{n+1}^\pi(h_n, d_n(h_n), \sigma) \\
=\ & r_n(s_n, d_n(h_n)) + E_{h_n}^\pi \left(u_{n+1}^\pi(h_n, d_n(h_n), X_{n+1}) \right) \\
=\ & r_n(s_n, d_n(h_n)) \\
& + E_{h_n}^\pi \left(E_{h_{n+1}}^\pi \left(\sum_{k=n+1}^{N-1} r_k(X_k, Y_k) + r_N(X_N) \right) \right) \\
=\ & r_n(s_n, d_n(h_n)) + E_{h_n}^\pi \left(\sum_{k=n+1}^{N-1} r_k(X_k, Y_k) + r_N(X_N) \right) \\
=\ & E_{h_n}^\pi \left(r_n(X_n, Y_n) + \sum_{k=n+1}^{N-1} r_k(X_k, Y_k) + r_N(X_N) \right) \\
=\ & E_{h_n}^\pi \left(\sum_{k=n}^{N-1} r_k(X_k, Y_k) + r_N(X_N) \right),
\end{aligned}
$$

which finishes the main part of the proof. As we have seen in (2.2) we also have $u_1^\pi(s) = v_N^\pi(s)$, so if we have an optimal policy we know the value of the underlying Markov decision problem. $\square$

We want to have a look at the complexity of the policy evaluation algorithm. The crucial point is that in every decision epoch we have to evaluate u_t^π for every possible history h_t. Let there be α states and β actions and N decision epochs. Then there are altogether $\alpha^N \beta^{N-1}$ possible histories since histories are of the form $h_N = (s_1, a_1, s_2, a_2, \ldots, a_{N-1}, s_N)$. Up to time t there are $\alpha^t \beta^{t-1}$ possible histories. Now fix a policy π, a decision epoch t and a particular history h_t up to time t. To calculate $u_t^\pi(h_t)$ Algorithm 1 needs $\alpha\beta$ multiplications, see line five. To construct u_t^π we need to evaluate $u_t^\pi(h_t)$ for every $h_t \in H_t$, so $\alpha\beta\alpha^t\beta^{t-1} = \alpha^{t+1}\beta^t$ multiplications are needed. Now we iterate over all decision epochs, so all together $\sum_{t=1}^{N-1} \alpha^{t+1}\beta^t$ multiplications are needed to evaluate a single policy. Additionally we have to store $\alpha^N \beta^{N-1}$ numbers in the beginning of the algorithm in order to construct u_N^π.

So this is quite a lot of work to do. Luckily we will see that we only have to give attention to deterministic Markovian decision rules. In the Markovian case u_t^π are actually functions from S to $\mathbb{R}$ since the actions chosen only depend on the current state s_t and not on the entire past h_t. Under this assumption we have

$$u_t^\pi(h_t) = E_{h_t}^\pi \left(\sum_{n=t}^{N-1} r_n(X_n, Y_n) + r_N(X_N) \right)$$
$$= E_{s_t}^\pi \left(\sum_{n=t}^{N-1} r_n(X_n, Y_n) + r_N(X_N) \right).$$

Consequently we do not have to look at each history h_t like we did in Algorithm 1. We rewrite the policy evaluation algorithm for the case $\pi \in \mathcal{D}^{DM}$. Now we have a look at line five of Algorithm 2. We need α multiplications for a single fixed s_t, so constructing u_t^π needs only α^2 multiplications. We again iterate over the entire time horizon which leads to $\sum_{t=1}^{N-1} \alpha^2 = (N-1)\alpha^2$ multiplications altogether. Additionally we only have to store α numbers to construct u_N^π. Theorem 2.3 also includes the correctness of Algorithm 2 if deterministic Markovian policies are interpreted as degenerated probability measures.

Algorithm 2 Finite Horizon Policy Evaluation Algorithm for $\pi \in \mathcal{D}^{DM}$

1: for $t = N$ set $u_N^\pi(s_N) = r_N(s_N) \quad \forall s_N \in S$.
2: $t \leftarrow N$
3: **for** $t \neq 1$ **do**
4: $\quad t \leftarrow t - 1$
5: $\quad u_t^\pi(s_t) = r_t(s_t, d(s_t)) + \sum_{\sigma \in S} p_t(\sigma | s_t, d_t(s_t)) u_{t+1}^\pi(\sigma)$ for all $s_t \in S$
6: **end for**

2.1.2 The Bellman Equations

Now we can take a look at the Bellman equations. The solution of the equations will help us to find optimal policies.

We define the functions

$$u_t^*(h_t) = \sup_{\pi \in \mathcal{D}^{RH}} u_t^\pi(h_t) \tag{2.3}$$

which are the best possible expected total rewards from decision epoch t onwards if the history until t equals h_t and the supremum is attained. This is, e.g., the case if we deal with finite S and A.

The Bellman equations, often also called optimality equations, are given by

$$u_t(h_t) = \sup_{a \in A_{s_t}} \left(r_t(s_t, a) + \sum_{\sigma \in S} p_t(\sigma | s_t, a) u_{t+1}(h_t, a, \sigma) \right) \tag{2.4a}$$

for $1 \leq t \leq N - 1$, and

$$u_N(h_N) = r_N(s_N). \tag{2.4b}$$

The solution of the Bellman equations is a sequence of functions $u_t : H_t \to \mathbb{R}$ fulfilling (2.4a) and (2.4b). We will prove that the solution of the Bellman equations fulfil (2.3), so if we solve the Bellman equations we obtain a finite sequence $u_t : H_t \to \mathbb{R}$ of functions that tell us what the best possible expected reward from decision epoch t

onwards is if the history up to time t equals h_t. Before we do this we need a small lemma.

Lemma 2.4. *Let f be a real–valued function on a discrete set Ω and let $p(\cdot)$ be a probability distribution on Ω. Then we have*

$$\sup_{\omega \in \Omega} f(\omega) \geq \sum_{\omega \in \Omega} p(\omega) f(\omega).$$

Proof. Set $\omega^* := \sup_{\omega \in \Omega} f(\omega)$. We easily calculate

$$\omega^* = \omega^* \sum_{\omega \in \Omega} p(\omega) = \sum_{\omega \in \Omega} p(\omega) \omega^* \geq \sum_{\omega \in \Omega} p(\omega) f(\omega).$$

$\square$

Now we can state the main property of the solution of the Bellman equations.

Theorem 2.5. *Suppose the family $u_t, t = 1, \ldots, N$ is a solution of the Bellman equations. Then we have*

$$u_t(h_t) = u_t^*(h_t)$$

for all $h_t \in H_t$ and $t = 1, \ldots, N$. Moreover we have $u_1(s) = v_N^*(s)$ for all $s \in S$, i.e., u_1 equals the value of the underlying Markov decision problem.

Proof. Within this proof we denote the solution[1] of the Bellman equations (assuming it exists) by $u_t, t = 1, \ldots, N$. We are starting with proving by backward induction that

$$u_t(h_t) \geq u_t^*(h_t) \quad \forall h_t \in H_t, \quad t = 1, \ldots, N. \tag{2.5}$$

Note that we have for an arbitrary policy π by definition $u_N^\pi(h_N) = E_{h_N}(r_N(X_N)) = r_N(s_N)$ since by conditioning on an arbitrary history

---

[1]If S and A are finite we only have to assume that $r_t(\cdot, \cdot)$ and $r_N(\cdot)$ are bounded, then a unique solution always exists.

$h_N \in H_N$ the random variable X_N is known. By (2.4b) we have $u_N(h_N) = r_N(s_N) = u_N^\pi(h_N)$ for all $h_N \in H_N$ and an arbitrary $\pi \in \mathcal{D}^{RH}$. So consequently we have $u_N(h_N) = u_N^*(h_N)$ for all $h_N \in H_N$ and of course therefore $u_N(h_N) \geq u_N^*(h_N)$ for all $h_N \in H_N$. Now assume that

$$u_t(h_t) \geq u_t^*(h_t) \quad \forall h_t \in H_t, \quad t = n+1, \dots, N,$$

and let $\tilde{\pi} := (\tilde{d}_1, \dots, \tilde{d}_N)$ be an arbitrary randomized history dependent policy. For $t = n$ we have

$$u_n(h_n) \overset{(2.4)}{=} \sup_{a \in A_{s_n}} \left(r_n(s_n, a) + \sum_{\sigma \in S} p_n(\sigma|s_n, a) u_{n+1}(h_n, a, \sigma) \right)$$

$$\overset{\text{i.h.}}{\geq} \sup_{a \in A_{s_n}} \left(r_n(s_n, a) + \sum_{\sigma \in S} p_n(\sigma|s_n, a) u_{n+1}^*(h_n, a, \sigma) \right)$$

$$\overset{(2.3)}{\geq} \sup_{a \in A_{s_n}} \left(r_n(s_n, a) + \sum_{\sigma \in S} p_n(\sigma|s_n, a) u_{n+1}^{\tilde{\pi}}(h_n, a, \sigma) \right)$$

$$\overset{2.4}{\geq} \sum_{a \in A} q_{\tilde{d}_n(h_n)}(a)$$

$$\left(r_n(s_n, a) + \sum_{\sigma \in S} p_n(\sigma|s_n, a) u_{n+1}^{\tilde{\pi}}(h_n, a, \sigma) \right)$$

$$\overset{(2.3)}{=} u_n^{\tilde{\pi}}(h_n).$$

Because $\tilde{\pi}$ was arbitrary we showed $u_t(h_t) \geq u_t^*(h_t)$ for all $h_t \in H_t$, $t = 1, \dots, N$. So we have proved the claim (2.5). Now we want to show that, for an arbitrary $\varepsilon > 0$, there exists a policy π for which we have

$$u_t^\pi(h_t) + (N - t)\varepsilon \geq u_t(h_t) \quad \forall h_t \in H_t, \quad t = 1, \dots, N. \qquad (2.6)$$

To do so we choose any policy[2] $\pi = (d_1, \ldots, d_{N-1})$ which fulfills for all $t = 1, \ldots, N$

$$r_t(s_t, d_t(h_t)) + \sum_{\sigma \in S} p_t(\sigma | s_t, d_t(h_t)) u_{t+1}(s_t, d_t(h_t), \sigma) + \varepsilon \geq u_t(h_t).$$

We again proof the claim (2.6) by backward induction. We have $u_N^\pi(h_N) = u_N(h_N)$ for an arbitrary policy π, so (2.6) clearly holds for $t = N$. Now assume that (2.6) is valid for all $t = n + 1, \ldots, N$. Then we have

$$u_n^\pi(h_n) = r_n(s_n, d_n(h_n)) + \sum_{\sigma \in S} p_n(\sigma | s_n, d_n(h_n)) u_{n+1}^\pi(s_n, d_n(h_n), \sigma)$$

$$\geq r_n(s_n, d_n(h_n))$$
$$+ \sum_{\sigma \in S} p_n(\sigma | s_n, d_n(h_n)) \left(u_{n+1}(h_n, d_n(h_n), \sigma) - (N - n - 1)\varepsilon \right)$$

$$= -(N - n)\varepsilon + r_n(s_n, d_n(h_n))$$
$$+ \sum_{\sigma \in S} p_n(\sigma | s_n, d_n(h_n)) u_{n+1}(h_n, d_n(h_n, \sigma)) + \varepsilon$$

$$\geq u_n(h_n) - (N - n)\varepsilon.$$

This proves the claim (2.6). By definition we have $u_t^*(\cdot) \geq u_t^\pi(\cdot)$ for all possible policies. Therefore we have

$$u_t^*(h_t) + (N - t)\varepsilon \geq u_t^\pi(h_t) + (N - t)\varepsilon \overset{(2.6)}{\geq} u_t(h_t) \overset{(2.5)}{\geq} u_t^*(h_t).$$

Now let us set $\tilde{\varepsilon} := (N - t)\varepsilon$. Then we have

$$u_t^*(h_t) + \tilde{\varepsilon} \geq u_t(h_t) \geq u_t^*(h_t),$$

which means $u_t^*(h_t) = u_t(h_t)$ since ε was arbitrary. Moreover because of $v_N^*(s) = u_1^*(s)$ we also have $u_1(s) = v_N^*(s)$. $\qquad\square$

---

[2] Such a policy clearly exists. Note that without adding ε we would have equality by definition if the image of h_t under d_t equals the best possible action.

Now we show how to use the solution of the Bellman equations to construct optimal policies. At first we take a look at the case when the Bellman equations attain the suprema.

Theorem 2.6. *Let $u_t^*, t = 1, \ldots, N$ be a finite sequence of functions which solve the Bellman equations (2.4a) and (2.4b) and assume that the policy $\mathcal{D}^{DH} \ni \pi^* := (d_1^*, \ldots, d_{N-1}^*)$ satisfies*

$$d_t^*(h_t) \in \operatorname*{argmax}_{a \in A_{s_t}} \; r_t(s_t, a) + \sum_{\sigma \in S} p_t(\sigma | s_t, a) u_{t+1}^*(h_t, a, \sigma). \qquad (2.7)$$

Then we have

$$u_t^{\pi^*}(h_t) = u_t^*(h_t), \;\; h_t \in H_t.$$

Moreover π^* is an optimal policy since we have $v_N^{\pi^*}(s) = v_N^*(s)$.

Remark 2.7 (Randomized vs. deterministic policies). Note that Theorem 2.6 is already stated for deterministic history dependent policies. This is not a restriction if seen in the following way: Let us rewrite (2.7) for randomized history dependent policies,

$$q_{d_t(h_t)}^*(\cdot) \in \operatorname*{argmax}_{q_{d_t(h_t)}(\cdot) \mathcal{P}(A_{s_t})} \; \mathcal{U}(q_{d_t(h_t)})$$

with

$$\mathcal{U}(q_{d_t}) := \sum_{a \in A_{s_t}} q_{d_t}(a) \left(r_t(s_t, a) + \sum_{\sigma \in S} p_t(\sigma | s_t, a) u_{t+1}^*(h_t, a, \sigma) \right).$$

We maximize over all possible probability distributions, which are by definition randomized policies, on A_{s_t}. Now let us assume that we have found an optimal probability distribution $q_{d_t(h_t)}^*$. Then we can guarantee the existence of a deterministic history dependent policy. To construct it set $f_{h_t}(a) := r_t(s_t, a) + \sum_{\sigma \in S} p_t(\sigma | s_t, a) u_{t+1}^*(h_t, a, \sigma)$ and $p(a) := q_{d_t(h_t)}(a)$ and use Lemma 2.4, which basically tells us that we are also doing well with a degenerated probability distribution $\tilde{q}_{d_t(h_t)}^*$. To construct $\tilde{q}_{d_t(h_t)}^*$ fix any $a^* \in \operatorname{argmax}_{a \in A_{s_t}} f_{h_t}(a)$ and define

$$\tilde{q}_{d_t(h_t)}^*(a) = \begin{cases} 1 & \text{if } a = a^* \\ 0 & \text{otherwise.} \end{cases}$$

Using $\tilde{q}^*$ is as good as using q^* according to Lemma 2.4. Note that $\tilde{q}^*$ is actually a deterministic history dependent policy as it is a degenerate probability measure on A_{s_t}. Now we have constructed a deterministic history dependent policy which is as good as the original randomized history dependent policy. So we can restrict our attention to $\pi \in \mathcal{D}^{DH}$.

Now we are ready to prove the theorem above.

Proof. We again prove the claim by backward induction. Clearly we have $u_N^{\pi^*}(h_N) = u_N^*(h_N)$ since this by definition holds true for every policy. Assume that

$$u_t^*(h_t) = u_t^{\pi^*}(h_t) \quad \forall h_t \in H_t$$

for all $t = n+1, \ldots, N$. Before we perform the induction step we need an auxiliary consideration. Condition (2.7) can equivalently be expressed by choosing the policy $\pi^* = (d_1^*, \ldots, d_{N-1}^*)$ in such a way that $d_t^*(h_t)$ solves the optimization problem

$$\max_{a \in A_{s_t}} r_t(s_t, a) + \sum_{\sigma \in S} p_t(\sigma | s_t, a) u_{t+1}^*(h_t, a, \sigma).$$

Note that we have

$$u_t^*(h_t) = \max_{a \in A_{s_t}} r_t(s_t, a) + \sum_{\sigma \in S} p_t(\sigma | s_t, a) u_{t+1}^*(h_t, a, \sigma)$$

by definition (2.4a) as well as

$$u_t^{\pi^*}(h_t) = r_t(s_t, d_t^*(h_t)) + \sum_{\sigma \in S} p_t(\sigma | s_t, d_t^*(h_t)) u_{t+1}^{\pi^*}(h_t, d_t^*(h_t), \sigma)$$

by Theorem 2.3. So using the the induction hypothesis $u_t^*(h_t) = u_t^{\pi^*}(h_t)$ yields

$$r_t(s_t, d_t^*(h_t)) + \sum_{\sigma \in S} p_t(\sigma | s_t, d_t^*(h_t)) u_{t+1}^{\pi^*}(h_t, d_t^*(h_t), \sigma)$$

$$= \max_{a \in A_{s_t}} r_t(s_t, a) + \sum_{\sigma \in S} p_t(\sigma | s_t, a) u_{t+1}^*(h_t, a, \sigma)$$

(2.8)

for all $t = n + 1, \ldots, N - 1$. Now for $h_n = (h_{n-1}, d^*_{n-1}(h_{n-1}), s_n)$ we calculate

$$
\begin{aligned}
u^*_n(h_n) &= \max_{a \in A_{s_n}} r_n(s_n, a) + \sum_{\sigma \in S} p_n(\sigma | s_n, a) u^*_{n+1}(h_n, a, \sigma) \\
&= r_n(s_n, d^*_n(h_n)) + \sum_{\sigma \in S} p_n(\sigma | s_n, d^*_n(h_n)) u^{\pi^*}_{n+1}(h_n, d^*_n(h_n), \sigma) \\
&= u^{\pi^*}_n(h_n).
\end{aligned}
$$

Indeed the constructed policy is optimal since $v^{\pi^*}_N(s) = u^{\pi^*}_1(s) = u^*_1(s) = v^*_N(s)$ for all possible starting states $s \in S$. $\square$

Theorem 2.6 asserts that there exists a deterministic history dependent policy which is optimal since it gives instruction how to construct it whenever, as assumed in the theorem, the supremum is attained, meaning there exists an $a^* \in A_{s_t}$ for which we have

$$
\begin{aligned}
r_t(s_t, a^*) + &\sum_{\sigma \in S} p_t(\sigma | s_t, a^*) u^*_{t+1}(h_t, a^*, \sigma) \\
&= \sup_{a \in A_{s_t}} r_t(s_t, a) + \sum_{\sigma \in S} p_t(\sigma | s_t, a) u^*_{t+1}(h_t, a, \sigma). \quad (2.9)
\end{aligned}
$$

Remark 2.8. Criteria which establish (2.9):

1. If A is finite, then (2.9) is fulfilled.

2. If A_s is compact for every $s \in S$, $r_t(s, a)$ and $p_t(\sigma | s, a)$ are continuous in a for every fixed $s \in S$ and additionally $r_t(s, a)$ and the final reward $r_N(s)$ are bounded functions then (2.9) is fulfilled as well.

If however (2.9) is not fulfilled then we only have ε–optimal policies.

Theorem 2.9. Fix $\varepsilon > 0$ and let the family u^*_t, $t = 1, \ldots, N$ be a solution of the Bellman equations. Furthermore choose $\pi = (d_1, \ldots, d_{N-1}) \in \mathcal{D}^{DH}$ such that

$$
\begin{aligned}
r_t(s_t, d_t(h_t)) + &\sum_{\sigma \in S} p_t(\sigma | s_t, d_t(h_t)) u^*_{t+1}(h_t, d_t(h_t), \sigma) + \frac{\varepsilon}{N-1} \\
&\geq \sup_{a \in A_{s_t}} r_t(s_t, a) + \sum_{\sigma \in S} p_t(\sigma | s_t, a) u^*_{t+1}(h_t, a, \sigma)
\end{aligned}
$$

for $t = 1, 2, \ldots, N - 1$. *Then the following statements are true:*

1. *For every* $t = 1, \ldots, N - 1$

$$u_t^\pi(h_t) + (N - t)\frac{\varepsilon}{N - 1} \geq u_t^*(h_t), \quad h_t \in H_t.$$

2. *The policy* π *is* ε*-optimal,* $v_N^\pi(s) + \varepsilon \geq v_N^*(s)$ *for all* $s \in S$.

Proof. 1. We have $u_N^\pi(h_n) = r_N(s_N) = u_N^*(s_N)$ for all

$$h_N = (h_{N-1}, a_{N-1}, s_N),$$

hence the inequality follows for $t = N$. Now assume that the statement is correct for $t = n + 1, \ldots, N$. Then we have

$$
\begin{aligned}
u_n^\pi(h_n) &= r_n(s_n, d_n(h_n)) \\
&\quad + \sum_{\sigma \in S} p_n(\sigma | s_n, d_n(h_n)) u_{n+1}^\pi(s_n, d_n(h_n), \sigma) \\
&\geq r_n(s_n, d_n(h_n)) \\
&\quad + \sum_{\sigma \in S} p_n(\sigma | s_n, d_n(h_n)) u_{n+1}^*(s_n, d_n(h_n), \sigma) \\
&\quad - \frac{\varepsilon(N - t - 1)}{N - 1} \\
&\geq \sup_{a \in A_{s_t}} r_n(s_n, a) + \sum_{\sigma \in S} p_n(\sigma | s_t, a) u_{n+1}^*(h_t, a, \sigma) \\
&\quad - (N - t)\frac{\varepsilon}{N - 1} \\
&= u_n^*(h_n) - (N - t)\frac{\varepsilon}{N - 1}.
\end{aligned}
$$

2. Choose $t = 1$, then it follows from the first part that

$$v_N^\pi(s) + \varepsilon = u_1^\pi(s) + \varepsilon \geq u_1^*(s) = v_N^*(s)$$

for all $s \in S$.

$\square$

We want to show now that it is actually enough to look at deterministic Markovian policies.

Theorem 2.10. *Let $u_t^* : H_t \to \mathbb{R}$ be a finite sequence of functions solving the Bellman equations (2.4a) and (2.4b). Then we have*

1. *For each $t = 1, \ldots, N$ the function $u_t^*(h_t)$ depends on the history h_t only through the current state s_t.*

2. *If condition (2.9) holds, i.e., the supremum is attained, there exists an optimal policy which is deterministic and Markovian.*

3. *Fix $\varepsilon > 0$, then there exists an ε-optimal deterministic Markovian policy.*

Proof. 1. We again use backward induction. We have $u_N^*(h_N) = u_N^*(h_{N-1}, a_{N-1}, s_N) = r_N(s_N)$ for all possible histories, so $u_N^*(h_N) = u_N^*(s_N)$ and therefore u_N^* only depends on the past through the current state. Now assume that $u_t^*(h_t) = u_t^*(s_t)$ for all $t = n+1, \ldots, N$. We have

$$u_n^*(h_n) = \sup_{a \in A_{s_n}} r_t(s_n, a) + \sum_{\sigma \in S} p_n(\sigma | s_n, a) u_{n+1}^*(h_n, a, \sigma)$$

$$= \sup_{a \in A_{s_n}} r_t(s_n, a) + \sum_{\sigma \in S} p_n(\sigma | s_n, a) u_{n+1}^*(\sigma). \qquad (2.10)$$

Now note that the term (2.10) depends on the history only through s_n, which means $u_n^*(h_n) = u_n^*(s_n)$ for all possible histories. This finishes the proof of the first part.

2. Now we know by (2.6) that choosing

$$d_t^*(h_t) \in \operatorname*{argmax}_{a \in A_{s_t}} r_t(s_t, a) + \sum_{\sigma \in S} p_t(\sigma | s_t, a) u_{t+1}^*(h_t, a, \sigma)$$

yields an optimal policy. We are rewriting this condition using our result (i), i.e., we replace $u_{t+1}^*(h_t, a, \sigma)$ by $u_{t+1}^*(\sigma)$,

$$d_t^*(h_t) \in \operatorname*{argmax}_{a \in A_{s_t}} r_t(s_t, a) + \sum_{\sigma \in S} p_t(\sigma | s_t, a) u_{t+1}^*(\sigma).$$

Algorithm 3 Backward induction algorithm for finite Markov decision problems

---

1: for $t = N$ set $u_N^*(s_N) = r_N(s_N)$ $\forall s_N \in S$.
2: $t \leftarrow N$
3: **while** $t \neq 1$ **do**
4:    $t \leftarrow t - 1$
5:    **for** $s_t \in S$ **do**
6:       Compute

$$u_t^*(s_t) = \max_{a \in A_{s_t}} r_t(s_t, a) + \sum_{\sigma \in S} p_t(\sigma|s_t, a) u_{t+1}^*(\sigma)$$

7:       Set

$$A_{s_t, t}^* = \operatorname*{argmax}_{a \in A_{s_t}} r_t(s_t, a) + \sum_{\sigma \in S} p_t(\sigma|s_t, a) u_{t+1}^*(\sigma)$$

8:    **end for**
9: **end while**

As we see d_t^* depends on the history only through the current state s_t so we actually have $d_t^*(h_t) = d_t^*(s_t)$. The policy $\pi^* = (d_1^*, \ldots, d_N^*)$ constructed by (2.6) is actually a deterministic Markovian policy.

3. Use Theorem 2.9, the rest is analogous to (ii).

$\square$

2.2 Backward Induction Algorithm

In this section we put everything together what we know about finite Markov decision problems and construct an algorithm which calculates an optimal deterministic Markovian policy. We again assume that the suprema in (2.9) are attained. We summarize our results in the following Theorem.

Theorem 2.11. *Let u_t^* and $A_{s_t,t}^*$ be constructed using Algorithm 3. Then the following statements are true:*

1. *For $t = 1, \ldots, N$ and $h_t = (h_{t-1}, a_{t-1}, s_t)$ we have*

$$u_t^*(s_t) = \sup_{\pi \in \mathcal{D}^{RH}} u_t^\pi(h_t).$$

2. *Let $d_t^*(s_t) \in A_{s_t,t}^*$ for all $s_t \in S$ and $t = 1, \ldots, N - 1$. Then $\pi^* = (d_1^*, \ldots, d_{N-1}^*)$ is an optimal deterministic Markovian policy and satisfies $v_N^{\pi^*}(s) = \sup_{\pi \in \mathcal{D}^{RH}} v_N^\pi(s)$ for all $s \in S$.*

Proof. We just need to summarize what we already know from previous theorems.

1. Looking at the algorithm reveals that u_t^* constructed by Algorithm 3 solves the Bellman equations. We showed in (2.5) that a property of the solution of the Bellman equations is $u_t^*(s_t) = \sup_{\pi \in \mathcal{D}^{RH}} u_t^\pi(h_t)$ which proves the first part.

2. Note that an $a \in A_{s_t,t}^*$ fulfils the property of Theorem 2.6. Therefore π^* is an optimal deterministic Markovian policy. Moreover we have by definition of the quantities, as already seen,

$$v_N^{\pi^*}(s) = u_1^{\pi^*}(s) = u_1^*(s) = \sup_{\pi \in \mathcal{D}^{RH}} u_1^\pi(s) = \sup_{\pi \in \mathcal{D}^{RH}} v_N^\pi(s).$$

$\square$

Now we should make some complexity considerations. In Algorithm 3 we have two nested loops. The outer one iterates over the decision epochs $N - 1, \ldots, 1$ and the inner one iterates over all $s \in S$. In every step we need to perform an optimization. This yields an effort of $O((N - 1)|S|\mathcal{M})$. Here $\mathcal{M}$ is the effort needed for solving the maximization step in the algorithm. If A is finite then we have a trivial worst case estimate for the optimization: We simply try out every $a \in A$ and store the a for which the objective function is maximal. Finding the maximum of a set of a values is linear and function evaluation is $O(S)$, so $\mathcal{M} = O(|S||A|)$ and the complexity of Algorithm 3 is $O((N-1)|S|^2|A|)$.

2.3 Examples

We now apply the solution techniques to the examples introduced in Section 1.3.

The (k, n) Card Game

In Section 1.3 we described the (k, n) card game which we now want to solve using the backward induction algorithm for the case $k = 10$ and $n = 20$. We take $T := n$, that leads to 19 decision epochs. $A := \{P, N\}$ where P stands for play and take another card and N stand for stop playing. We define the set of all states as $S := \{1, \ldots, kn\}$. Rewards and transition probabilities are defined as in Section 1.3. The optimal policy is stored in a matrix $\Pi \in \{P, N\}^{kn \times T-1}$. If we want to know what to do if we are in state i at time point j we have to look at Π_{ij}. The vector (2.11) is a part (first 20 entries) of a column of the policy as constructed by the backward induction algorithm. All the columns of Π are equal: we have $\Pi_{:,j_1} = \Pi_{:,j_2}$ for all $j_1, j_2 \in \{1, \ldots, 19\}$ since transition probabilities are stationary and intermediate rewards are all equal to zero. It is only important that we stop if $s > 13$. Time information does not play a role here.

$$(P\,P\,P\,P\,P\,P\,P\,P\,P\,P\,P\,P\,N\,N\,N\,N\,N\,N\,N\,N). \qquad (2.11)$$

We just need the first 20 entries because if we are in state $s > 20$ then action P as well as action N only leads to a reward of $R = 0$, so they are equally bad. We see that we have twelve consecutive p. This means that if the sum fulfils $s \leq 12$ we should take another card. If $s \geq 13$ we should stop playing.

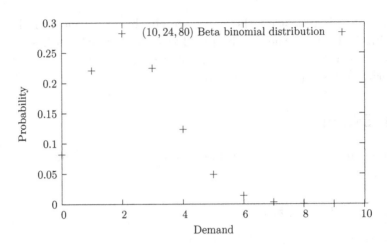

Figure 2.1: Beta binomial distribution

Stochastic Inventory Control

We set up our model with $S = \{0, \ldots, 10\}$, $A_s = \{0, \ldots, 10 - s\}$ and $T = 10$. Let the demand D_t be $(10, 24, 80)$ beta binomial distributed for all $t \in \{1, \ldots, 9\}$. This distribution seems a plausible choice for the demand D_t as can be seen in Figure 2.1. Based on the beta binomial distribution we can set up the transition probabilities as described in Section 1.3. We further choose $f(j) = 50j$, $h(s + a) = s + a$ and

$$b(a) = \begin{cases} 20 + 35a & \text{if } a > 0 \\ 0 & \text{otherwise.} \end{cases}$$

With this information we can set up our rewards $r(s, a) = F(s + a) - b(a) - h(s + a)$ where $F(s + a)$ is the expected revenue if $s + a$ units are stored in the warehouse, $b(a)$ are the costs for ordering a units and $h(s + a)$ is the cost for maintaining $s + a$ units in the warehouse. We define the terminal rewards $r_{10}(s) := 0$ for all $s \in S$. The optimal policy is then given by (2.12) where Π_{ij} contains the information what action to choose at time point j when currently i units are stored in

$$\Pi = \begin{pmatrix} 10 & 10 & 10 & 10 & 8 & 7 & 5 & 3 & 0 \\ 9 & 9 & 9 & 9 & 7 & 6 & 4 & 0 & 0 \\ 0 & 0 & 0 & 8 & 6 & 5 & 0 & 0 & 0 \\ 0 & 0 & 0 & 0 & 0 & 0 & 0 & 0 & 0 \\ 0 & 0 & 0 & 0 & 0 & 0 & 0 & 0 & 0 \\ 0 & 0 & 0 & 0 & 0 & 0 & 0 & 0 & 0 \\ 0 & 0 & 0 & 0 & 0 & 0 & 0 & 0 & 0 \\ 0 & 0 & 0 & 0 & 0 & 0 & 0 & 0 & 0 \\ 0 & 0 & 0 & 0 & 0 & 0 & 0 & 0 & 0 \\ 0 & 0 & 0 & 0 & 0 & 0 & 0 & 0 & 0 \\ 0 & 0 & 0 & 0 & 0 & 0 & 0 & 0 & 0 \end{pmatrix}. \tag{2.12}$$

the warehouse. So Π_{ij} tells us how many additional units have to be bought to act optimal.

3 Infinite Horizon Markov Decision Problems

We consider infinite horizon Markov decision problems. We state the Bellman equations which characterize optimal policies. They are an important tool for proving the optimality of so called stationary policies. Then we take a look at two important algorithms which solve infinite Markov decision problems: Value Iteration and Policy Iteration. In this chapter we follow the book of [Put94]. Furthermore we use the books [Whi93] and [BR11].

In this chapter we take a look at the infinite horizon case. We introduce a discount factor $\lambda \in [0, 1)$ and define the *expected total discounted reward* of policy π as

$$v_\lambda^\pi(s) := \lim_{N \to \infty} E_s^\pi \left(\sum_{t=1}^N \lambda^{t-1} r(X_t, Y_t) \right). \tag{3.1}$$

If $\sup_{s \in S, a \in A} |r(s, a)| \leq c \in \mathbb{R}$ holds, the limit above exists, so we want to assume bounded rewards. In contrast to the finite horizon case we additionally assume time independent rewards and transition probabilities. Discounting arises naturally in many applications since a cash flow received later in time is often less worth than money which is earned right away. In this chapter it is always assumed that $\lambda \in (0, 1]$, this will play a key role in analysing infinite horizon Markov decision problems.

Definition 3.1. *The quintuple*

$$\mathcal{M} = (S, A, p(s'|s, a), r(s, a), \lambda)$$

together with an optimality criterion is called an infinite horizon Markov decision problem. Here we use the expected total discounted reward *criterion.*

We again call a policy π^* optimal if

$$v^{\pi^*}(s) \geq v^{\pi}(s)$$

for all starting states $s \in S$ and every policy $\pi \in \mathcal{D}^{RH}$. We define the value of an MDP as $v^*(s) = \sup_{\pi \in \mathcal{D}^{RH}} v^{\pi}(s)$. An optimal policy exists whenever $v^{\pi^*}(s) = v^*(s)$ for all $s \in S$.

Like in the finite horizon case we can actually restrict our attention to a smaller class of policies.

Theorem 3.2. *For any $\pi \in \mathcal{D}^{RH}$ there exists for each $s \in S$ a policy $\tilde{\pi} \in \mathcal{D}^{RM}$ for which $v_{\lambda}^{\tilde{\pi}}(s) = v_{\lambda}^{\pi}(s)$.*

Proof. We proof the claim by showing that for every $\pi \in \mathcal{D}^{RH}$ and a given $s \in S$ there exists a policy $\tilde{\pi} \in \mathcal{D}^{RM}$ for which we have

$$P^{\tilde{\pi}}(X_t = \sigma, Y_t = a | X_1 = s) = P^{\pi}(X_t = \sigma, Y_t = a | X_1 = s). \quad (3.2)$$

Observe furthermore that then

$$v_{\lambda}^{\pi}(s) = \sum_{t=1}^{\infty} \sum_{\sigma \in S} \sum_{a \in A_{\sigma}} \lambda^{t-1} r(\sigma, a) P^{\pi}(X_t = \sigma, Y_t = a | X_1 = s)$$

$$= \sum_{t=1}^{\infty} \sum_{\sigma \in S} \sum_{a \in A_{\sigma}} \lambda^{t-1} r(\sigma, a) P^{\tilde{\pi}}(X_t = \sigma, Y_t = a | X_1 = s)$$

$$= v_{\lambda}^{\tilde{\pi}}(s),$$

what would prove the claim. To establish (3.2) fix $s \in S$ and define the randomized Markovian policy $\tilde{\pi} = (\tilde{d}_1, \tilde{d}_2, \ldots)$ by

$$q_{\tilde{d}_t(\sigma)}(a) = P^{\pi}(Y_t = a | X_t = \sigma, X_1 = s).$$

This leads to

$$P^{\tilde{\pi}}(Y_t = a | X_t = \sigma) = P^{\tilde{\pi}}(Y_t = a | X_t = \sigma, X_1 = s)$$

$$= P^\pi(Y_t = a | X_t = \sigma, X_1 = s). \quad (3.3)$$

Now for $t = 1$ (3.2) is satisfied since for $s = \sigma$

$$P^{\tilde\pi}(X_1 = s, Y_1 = a | X_1 = s) = P^{\tilde\pi}(Y_1 = a | X_1 = s)$$
$$\stackrel{(3.3)}{=} P^\pi(Y_1 = a | X_1 = s)$$
$$= P^\pi(X_1 = s, Y_1 = a | X_1 = s).$$

Now assume that (3.2) is true for $t = 1, \ldots, n-1$ then

$$P^\pi(X_n = \sigma | X_1 = s)$$
$$= \sum_{k \in S} \sum_{a \in A_k} P^\pi(X_{n-1} = k, Y_{n-1} = a | X_1 = s) p(\sigma | k, a)$$
$$= \sum_{k \in S} \sum_{a \in A_k} P^{\tilde\pi}(X_{n-1} = k, Y_{n-1} = a | X_1 = s) p(\sigma | k, a)$$
$$= P^{\tilde\pi}(X_n = \sigma | x_1 = s).$$

Finally, we calculate

$$P^{\tilde\pi}(X_n = \sigma, Y_n = a | X_1 = s)$$
$$= P^{\tilde\pi}(Y_n = a | X_n = \sigma) P^{\tilde\pi}(X_n = \sigma | X_1 = s)$$
$$= P^\pi(Y_n = a | X_n = \sigma, X_1 = s) P^\pi(X_n = \sigma | X_1 = s)$$
$$= P^\pi(X_n = \sigma, Y_n = a | X_1 = s).$$

$\square$

Now we are introducing some vector based notation, which we use in the rest of the chapter. Let V be the set of bounded real valued functions on S equipped with the supremum norm $\|v\|_\infty = \sup_{s \in S} |v(s)|$. Let A be a linear operator on V. The norm $\| \cdot \|_\infty$ induces an operator norm $\|A\|_\infty = \sup_{\|v\|_\infty \leq 1} \|Av\|_\infty$.

For finite S elements in V are vectors and linear operators are matrices. In that case the induced matrix norm is given by

$$\|A\| = \sup_{s \in S} \sum_{j \in S} |A_{sj}|.$$

We want to abbreviate our notation a little in order to derive a more compact form of (3.1), therefore we define for any decision rule $d \in D^{DM}$

$$r_d(s) := r(s, d(s)) \text{ and } p_d(j|s) := p(j|s, d(s))$$

and for $d \in D^{RM}$

$$r_d(s) := \sum_{a \in A_s} q_{d(s)}(a) r(s, a) \text{ and } p_d(j|s) := \sum_{a \in A_s} q_{d(s)}(a) p(j|s, a).$$

We call r_d the reward vector (sequence if S is countable infinite) and P_d the transition probability matrix (operator acting on sequences) corresponding to the decision rule d where

$$(P_d)_{sj} := p_d(j|s). \tag{3.4}$$

Lemma 3.3. *Let $v \in V$ be arbitrary, then it follows that $r_d + \lambda P_d v \in V$.*

Proof. Because we have assumed that the rewards are bounded we have $\|r_d\| \le c$ for an arbitrary decision rule d. Furthermore we have $\|P_d v\| \le \|P_d\| \|v\| = \|v\|$ because $\|P_d\| = \sup_{\|v\|_\infty \le 1} \|P_d v\| = 1$. So altogether we have

$$\|r_d + \lambda P_d v\| \le \|r_d\| + |\lambda| \|P_d v\| \le \|r_d\| + \|v\| = \tilde{c}.$$

$\square$

Now fix a policy $\pi = (d_1, d_2, \dots) \in \mathcal{D}^{RM}$. We calculate the t-step transition probability under the policy π,

$$p_\pi^t(j|s) := P^\pi(X_{t+1} = j | X_1 = s) = (P_{d_t} \cdots P_{d_1})_{sj},$$

where P_{d_i} is defined in (3.4). Now we again build a matrix using the t-step transition probabilities,

$$(P_\pi^t)_{sj} := p_\pi^t(j|s). \tag{3.5}$$

Using this notation we have

$$
\begin{aligned}
v_\lambda^\pi(s) &= \lim_{N\to\infty} E_s^\pi \left(\sum_{t=1}^{N} \lambda^{t-1} r(X_t, Y_t) \right) \\
&= \lim_{N\to\infty} \sum_{t=1}^{N} \lambda^{t-1} E_s^\pi \left(r(X_t, Y_t) \right) \\
&= \lim_{N\to\infty} \sum_{t=1}^{N} \lambda^{t-1} \left(P_\pi^{t-1} \right)_{s:} r_{d_t} \\
&= \sum_{t=1}^{\infty} \lambda^{t-1} \left(P_\pi^{t-1} \right)_{s:} r_{d_t},
\end{aligned}
$$

where we set $P_\pi^0 := \mathrm{id}_V$ to include the immediate first reward which only depends on the initial state s_1 and the first action a_1. This leads to a more compact representation of the expected discounted reward which is our first formula for policy evaluation.

Lemma 3.4. *Let π be an arbitrary Markovian randomized (which includes the deterministic case) policy. Then the expected total discounted reward fulfils*

$$
v_\lambda^\pi = \sum_{t=1}^{\infty} \lambda^{t-1} P_\pi^{t-1} r_{d_t}.
$$

Proof. The calculation above proves the claim.   □

3.1 Bellman Equations and Existence of Optimal Policies

We will start with an alternative formula for evaluating so called stationary policies. A stationary policy uses the same decision rule in every decision epoch, $\Delta = (d, d, \dots)$. We will see that these policies play an important role in infinite horizon models.

First of all we need some results from functional analysis. An operator $A : V \to V$ is called bounded if $\|Av\| \leq c\|v\|$ for all $v \in V$ and a fixed $c \in \mathbb{R}$. We define the *spectral radius* of a bounded linear operator $L : V \to V$ by $\rho(L) := \lim_{n \to \infty} \|L^n\|^{\frac{1}{n}}$.

Lemma 3.5. *Let $L : V \to V$ be a bounded linear operator. Then we have that*

1. $\rho(L) \leq \|L\|$.

2. *If $\rho(L) < 1$ then $(id_V - L)^{-1}$ exists.*

An operator $A : V \to V$ is called a contraction mapping if there exists a $\lambda \in [0, 1)$ such that $\|Av - Au\| \leq \lambda\|v - u\|$.

Theorem 3.6 (Banach's Fixed-Point Theorem). *Let V be a Banach space and $A : V \to V$ be a contraction mapping. Then the following holds true:*

1. *The equation $Av = v$ has a unique solution $v^* \in V$.*

2. *Let $v_0 \in V$ be arbitrary then the sequence $(v_n)_{n \in \mathbb{N}}$ defined by*

$$v_{n+1} := Av_n$$

converges to v^*.

Now we are ready to prove a first theoretically important result. It will help us to identify so called stationary policies to be optimal. A stationary policy is of the form $\Delta = (d, d, \ldots)$, so it uses the same decision rule in every decision epoch. In the following proof we will already illustrate how the Banach fixed-point theorem is related to infinite Markov decision problems although the next result can be verified without using it. Later the Banach fixed-point theorem will become very important.

Theorem 3.7. *Let Δ be a stationary policy in $\mathcal{D}^{RM}$. Then the total expected discounted reward v_λ^Δ is the unique solution of $v = r_d + \lambda P_d v$ in V. Furthermore, we have*

$$v_\lambda^\Delta = (id_V - \lambda P_d)^{-1} r_d = \sum_{t=1}^{\infty} \lambda^{t-1} P_d^{t-1} r_d.$$

Proof. First we note that by Lemma 3.4 we have for an arbitrary $\pi = (d_1, d_2, \ldots) \in \mathcal{D}^{RM}$

$$
\begin{aligned}
v_\lambda^\pi &= \sum_{t=1}^\infty \lambda^{t-1} P_\pi^{t-1} r_{d_t} \\
&= r_{d_1} + \lambda P_{d_1} r_{d_2} + \lambda^2 P_{d_1} P_{d_2} r_{d_3} + \cdots \\
&= r_{d_1} + \lambda P_{d_1} \left(r_{d_2} + \lambda P_{d_2} r_{d_3} + \lambda^2 P_{d_2} P_{d_3} r_{d_4} + \cdots \right) \\
&= r_{d_1} + \lambda P_{d_1} v_\lambda^{\tilde{\pi}},
\end{aligned}
$$

where $\tilde{\pi} := (d_2, d_3, \ldots)$. Now we apply the fact above to the stationary decision rule $\Delta = (d, d, \ldots)$ and obtain

$$
v_\lambda^\Delta = r_d + \lambda P_d v_\lambda^\Delta, \tag{3.6}
$$

so v_λ^Δ solves the equation $v = r_d + \lambda P_d v$.

Next we have to show that v_λ^Δ is unique and an element of V. To do so define the linear transformation $L_d : V \to V$ given by $L_d v = r_d + \lambda P_d v$. Lemma 3.3 asserts that L_d indeed maps bounded functions to bounded functions. Now we have $L_d v_\lambda^\Delta = v_\lambda^\Delta$ which means v_λ^Δ is a fixed point of L_d. Furthermore, L_d is a contraction since it fulfils a Lipschitz condition

$$
\|L_d v - L_d w\| = \|\lambda P_d v - \lambda P_d w\| \le |\lambda| \|P_d\| \|v - w\| = |\lambda| \|v - w\|,
$$

where we assumed $\lambda < 1$. Because $(V, \|\cdot\|_\infty)$ is complete $v_\lambda^\Delta \in V$ as well as uniqueness of the solution follow by Banach's fixed point theorem.

We furthermore know by rewriting (3.6) that $(\mathrm{id}_V - \lambda P_d) v_\lambda^\Delta = r_d$. Now $(\mathrm{id}_V - \lambda P_d)^{-1}$ exists because of Lemma 3.5 since $\|\lambda P_d\| \le |\lambda| \|P_d\| = \lambda < 1$. Therefore we have

$$
(\mathrm{id}_V - \lambda P_d)^{-1} r_d = v_\lambda^\Delta = \sum_{t=1}^\infty \lambda^{t-1} P_d^{t-1} r_d,
$$

where the last equality is satisfied because of Lemma 3.4. $\square$

Definition 3.8 (Bellman equations, Optimality equations). *The equations*

$$v(s) = \sup_{a \in A_s} r(s, a) + \sum_{\sigma \in S} \lambda p(\sigma | s, a) v(\sigma)$$

are called Bellman equations or optimality equations. We can use our vector based notation to rewrite these equations as

$$\mathcal{B}v = v,$$

where the operator $\mathcal{B} : V \to V$ is defined by

$$\mathcal{B}v := \sup_{d \in D^{DM}} r_d + \lambda P_d v.$$

We further define $L_d : V \to V$ by

$$L_d v := r_d + P_d v.$$

From the proof of Theorem 3.7 we already know that L_d is a contraction mapping.

Remark 3.9. 1. Recall that $d \in D^{DM}$ is a function $d : S \to A$. The set S is assumed to be discrete, so d can be regarded as a vector in the finite case or as a sequence in the infinite case. The supremum above is computed with respect to the componentwise partial ordering, $v \leq u :\Leftrightarrow v_i \leq u_i$.

2. For all $v \in V$ we have $\sup_{d \in D^{RM}} r_d + \lambda P_d v = \sup_{d \in D^{DM}} r_d + \lambda P_d v$. Since $D^{RM} \supseteq D^{DM}$ we have $\sup_{d \in D^{RM}} r_d + \lambda P_d v \geq \sup_{d \in D^{DM}} r_d + \lambda P_d v$. To establish the reverse inequality choose $v \in V$ and $\tilde{d} \in D^{RM}$ and use Lemma 2.4 for each $s \in S$ with $\Omega = A_s$, $p = q_{\tilde{d}}$ and

$$f(\cdot) = r(s, \cdot) + \sum_{\sigma \in S} \lambda p(\sigma | s, \cdot) v(\sigma)$$

to obtain

$$\sup_{a \in A_s} r(s, a) + \sum_{\sigma \in S} \lambda p(\sigma | s, a) v(\sigma)$$

$$\geq \sum_{a \in A_s} q_{\tilde{d}}(a) \left(r(s,a) + \sum_{\sigma \in S} \lambda p(\sigma|s,a) v(\sigma) \right).$$

Hence we have

$$\sup_{d \in D^{MD}} r_d + \lambda P_d v \geq r_{\tilde{d}} + \lambda P_{\tilde{d}} v.$$

Because $\tilde{d} \in D^{RM}$ was arbitrary we have established the reverse inequality and thus equality.

3. If the action set is finite the supremum is trivially attained.

4. Solutions of the Bellman equations are fixed points of $\mathcal{B}$.

The following theorem states that the solution of the Bellman equations, provided it exists, equals the value of the underlying infinite Markov decision problem. Beforehand we need a small lemma.

Lemma 3.10. *Let $d \in D^{RM}$ be an arbitrary randomized decision rule.*

1. If $u, v \in V$ such that $u \geq v$ we have

$$(id_V - \lambda P_d)^{-1} u \geq (id_V - \lambda P_d)^{-1} v.$$

2. If $V \ni x \geq 0$ we have $(id_V - \lambda P_d)^{-1} x \geq 0$ and $(id_V - \lambda P_d)^{-1} x \geq x$.

Proof. We start by proving the second part. We again have $\rho(\lambda P_d) < 1$ so $(\mathrm{id}_V - \lambda P_d)^{-1}$ exists. Let $x \geq 0$. We apply Theorem 3.7 with $r_d = x$ and obtain

$$(\mathrm{id}_V - \lambda P_d)^{-1} x = \sum_{t=1}^{\infty} \lambda^{t-1} P_\pi^{t-1} x = x + \sum_{t=2}^{\infty} \lambda^{t-1} P_\pi^{t-1} x \geq x \geq 0$$

hence the second statement follows. For the first statement set $x := u - v$ and use the inequality above,

$$(\mathrm{id}_V - \lambda P_d)^{-1} (u - v) \geq 0.$$

Now $(\mathrm{id}_V - \lambda P_d)^{-1}$ is linear, this is clear if S is finite, so the claim follows. Nevertheless linearity also follows from the fact that $(\mathrm{id}_V - \lambda P_d)^{-1}(u-v) = \sum_{t=1}^{\infty} \lambda^{t-1} P_\pi^{t-1}(u-v)$ where the sum is absolutely convergent. Therefore the order of summation can be interchanged and we have $\sum_{t=1}^{\infty} \lambda^{t-1} P_\pi^{t-1}(u-v) = \sum_{t=1}^{\infty} \lambda^{t-1} P_\pi^{t-1} u - \sum_{t=1}^{\infty} \lambda^{t-1} P_\pi^{t-1} v = (\mathrm{id}_V - \lambda P_d)^{-1} u - (\mathrm{id}_V - \lambda P_d)^{-1} v$. Therefore, we can also include more general cases, e.g., if S is countably infinite, so the claim especially holds for any discrete set S. $\qquad\square$

Theorem 3.11. *Assume that there exists an element $v \in V$ with $\mathcal{B}v = v$. Then we have $v = v_\lambda^*$ and the solution v is unique.*

Proof. We will show the claim by establishing the following two statements:

1. If $\mathcal{B}v \leq v$ then $v_\lambda^* \leq v$.

2. If $\mathcal{B}v \geq v$ then $v_\lambda^* \geq v$.

Consequently if there exists $v \in V$ for which $\mathcal{B}v = v$ is fulfilled both statements above are true. That means $v_\lambda^* \leq v$ as well as $v_\lambda^* \geq v$ are satisfied, so $v = v_\lambda^*$.

We start with (i). Let $\pi = (d_1, d_2, \ldots) \in \mathcal{D}^{RM}$ be an arbitrary policy. Assume we have $v \in V$ with $v \geq \mathcal{B}v$. This means that

$$v \geq \sup_{d \in D^{DM}} r_d + \lambda P_d v = \sup_{d \in D^{RM}} r_d + \lambda P_d v, \qquad (3.7)$$

where the last equality is due to Remark 3.9. Therefore we have

$$v \geq r_{d_1} + \lambda P_{d_1} v \geq r_{d_1} + \lambda P_{d_1}(r_{d_2} + \lambda P_{d_2} v),$$

where we used (3.7) twice, the first time for decision rule d_1 and the second time for decision rule d_2. Using the notation introduced in (3.5) we simplify

$$r_{d_1} + \lambda P_{d_1}(r_{d_2} + \lambda P_{d_2} v) = r_{d_1} + \lambda P_{d_1} r_{d_2} + \lambda^2 P_{d_1} P_{d_2} v$$
$$= P_\pi^0 r_{d_1} + \lambda P_\pi^1 r_{d_2} + \lambda^2 P_\pi^2 v.$$

Iterating the above argument n times yields

$$v \geq \sum_{k=0}^{n-1} \lambda^k P_\pi^k r_{d_{k+1}} + \lambda^n P_\pi^n v.$$

Now we subtract $v_\lambda^\pi = \sum_{k=0}^\infty \lambda^k P_\pi^k r_{d_{k+1}}$ (Lemma 3.4) on both sides of the inequality above leading to

$$v - v_\lambda^\pi \geq \lambda^n P_\pi^n v - \sum_{k=n}^\infty \lambda^k P_\pi^k r_{d_{k+1}}. \tag{3.8}$$

Next we will show that the right hand side gets arbitrary small if n is chosen big enough. Let $\varepsilon > 0$ be arbitrary and define $\mathbf{1} := (1, \ldots, 1)$ or, if S is countable infinite $\mathbf{1} := (1, 1, \ldots)$ respectively.

Since $\|\lambda^n P_\pi^n v\| \leq \lambda^n \|v\|$ and $\lambda \in [0, 1)$ we have

$$-\varepsilon \mathbf{1} \leq \lambda^n P_\pi^n v \leq \varepsilon \mathbf{1}$$

if n is sufficiently large. Since rewards are assumed to be bounded, $r(s, a) \leq c$ for all $s \in S$ and $a \in A$ as well as $P_\pi^k \mathbf{1} = \mathbf{1}$, since P_π^k is the k-step transition matrix (operator if S is infinite), we have

$$\sum_{k=n}^\infty \lambda^k P_\pi^k r_{d_{k+1}} \leq c \sum_{k=n}^\infty \lambda^k \mathbf{1} = c\lambda^n \sum_{k=n}^\infty \lambda^{k-n} \mathbf{1}$$

$$= c\lambda^n \sum_{k=0}^\infty \lambda^k \mathbf{1}$$

$$\leq \frac{c\lambda^n}{1 - \lambda} \mathbf{1}$$

$$\leq \varepsilon \mathbf{1}$$

for n big enough and hence $-\sum_{k=n}^\infty \lambda^k P_\pi^k r_{d_{k+1}} \geq \varepsilon \mathbf{1}$. Combining these two estimates with (3.8) yields $v - v_\lambda^\pi \geq 2\varepsilon \mathbf{1}$ and therefore

$$v \geq v_\lambda^\pi + 2\varepsilon \mathbf{1}.$$

Now since ε as well as policy $\pi \in \mathcal{D}^{RM}$ have been arbitrary we have

$$v \geq \sup_{\pi \in \mathcal{D}^{RM}} v_\lambda^\pi = \sup_{\pi \in \mathcal{D}^{RH}} v_\lambda^\pi = v_\lambda^*,$$

where the first equality above is due to Theorem 3.2. Now we want to prove (ii). So assume $\mathcal{B}v \geq v$. Therefore we have $v \leq \sup_{d \in D^{DM}} r_d + \lambda P_d v$. Now choose $\varepsilon > 0$ arbitrary. There clearly exists a $d \in D^{DM}$ for which we have $v \leq r_d + \lambda P_d v + \varepsilon \mathbf{1}$ and therefore $r_d + \varepsilon \mathbf{1} \geq v - \lambda P_d v$. Now we are using Lemma 3.10 and obtain

$$\begin{aligned}
(\mathrm{id}_V - \lambda P_d)^{-1}(r_d + \varepsilon \mathbf{1}) &\geq (\mathrm{id}_V - \lambda P_d)^{-1}(v - \lambda P_d v) \\
&= (\mathrm{id}_V - \lambda P_d)^{-1}(\mathrm{id}_V - \lambda P_d)v \\
&= v.
\end{aligned}$$

We evaluate the left hand side of the inequality above,

$$\begin{aligned}
(\mathrm{id}_V - \lambda P_d)^{-1}(r_d + \varepsilon \mathbf{1}) &= v_\lambda^\Delta + \sum_{t=1}^{\infty} \lambda^{t-1} P_d^{t-1} \varepsilon \mathbf{1} \\
&= v_\lambda^\Delta + \varepsilon \sum_{t=1}^{\infty} \lambda^{t-1} \mathbf{1} \\
&= v_\lambda^\Delta + \frac{\varepsilon}{1-\lambda} \mathbf{1},
\end{aligned}$$

where the first equality is due to Theorem 3.7 and Lemma 3.4. Now we found a policy, namely the stationary policy $\Delta = (d, d, \ldots)$ for which we have

$$v \leq v_\lambda^\Delta + \frac{\varepsilon}{1-\lambda} \mathbf{1}.$$

Since ε was arbitrary we certainly have

$$v \leq \sup_{\pi \in \mathcal{D}^{RH}} v_\lambda^\pi = v_\lambda^*.$$

$\square$

Recall that the Bellman equations are of the form $\mathcal{B}v = v$. So we are looking for a fixed point, therefore we want to apply Banach's fixed-point theorem. This would ensure the existence of a solution of the Bellman equations. To do so we need to show that $\mathcal{B} : V \to V$ is a contraction mapping.

Lemma 3.12. *The Operator $\mathcal{B} : V \to V$ is a contraction mapping.*

Proof. Let $v \in V$ be arbitrary. $\|\mathcal{B}v\| = \|\sup_{d \in D^{DM}} r_d + \lambda P_d v\| = \sup_{d \in D^{DM}} \|r_d + \lambda P_d v\| \leq \sup_{d \in D^{DM}} \|r_d\| + \lambda \|P_d\| \|v\| \leq c + \lambda \|v\| \leq \tilde{c}$, hence $\mathcal{B} : V \to V$.

If $u = v$ the inequality $\|\mathcal{B}v - \mathcal{B}u\| \leq K \|u - v\|$ is fulfilled trivially for every $K \in \mathbb{R}$, so let $u \neq v$. Now choose $\varepsilon > 0$ so that

$$\sqrt{\varepsilon} < \|v - u\|.$$

Observe that this yields $\varepsilon < \sqrt{\varepsilon} \|v - u\|$. We will use this fact later. We want to show that

$$\|\mathcal{B}v - \mathcal{B}u\| \leq \lambda \|v - u\|$$

for $0 \leq \lambda < 1$. We are doing this componentwise by proving $|\mathcal{B}[v](s) - \mathcal{B}[u](s)| \lambda \leq \|v - u\|$. Now fix $s \in S$ and assume w.l.o.g. $\mathcal{B}[v](s) \geq \mathcal{B}[u](s)$. Then choose $d^*(s)$ in such a way that

$$r_{d^*}(s) + \lambda P_{d^*}[v](s) + \varepsilon \geq \sup_{d \in D^{DM}} r_d(s) + \lambda P_d[v](s). \tag{3.9}$$

Now we clearly have $\mathcal{B}[u](s) = \sup_{d \in D^{DM}} r_d + \lambda P_d u \geq r_{d^*} + \lambda P_{d^*} u$. We calculate, using $\|P_{d^*}\| = 1$ that

$$
\begin{aligned}
|\mathcal{B}[v](s) - \mathcal{B}[u](s)| &= \mathcal{B}[v](s) - \mathcal{B}[u](s) \\
&\leq \varepsilon + r_{d^*}(s) + \lambda P_{d^*}[v](s) - r_{d^*}(s) - \lambda P_{d^*}[u](s) \\
&\leq \varepsilon + \lambda |P_{d^*}[v - u](s)| \\
&\leq \varepsilon + \lambda \|P_{d^*}(v - u)\| \\
&\leq \varepsilon + \lambda \|v - u\| \\
&< \sqrt{\varepsilon} \|v - u\| + \lambda \|v - u\|
\end{aligned}
$$

$$= (\lambda + \sqrt{\varepsilon})\|v - u\|.$$

Now if $\mathcal{B}[v](s) < \mathcal{B}[u](s)$ then we also know $|\mathcal{B}[v](s) - \mathcal{B}[u](s)| = \mathcal{B}[u](s) - \mathcal{B}[v](s) < (\lambda + \varepsilon)\|v - u\|$. To see the last inequality the roles of v and u have to be interchanged in the argumentation above, beginning from inequality (3.9). Since $\varepsilon > 0$ was arbitrary we have $|\mathcal{B}[v](s) - \mathcal{B}[u](s)| \leq \lambda\|v - u\|$ for all $s \in S$. Therefore $\|\mathcal{B}v - \mathcal{B}u\| \leq \lambda\|v - u\|$ is satisfied for $0 \leq \lambda < 1$ by assumption, hence $\mathcal{B} : V \to V$ is a contraction mapping. $\qquad\square$

Now putting everything together what we have so far yields a central result for infinite horizon Markov decision problems. Since it is a main result we exactly recapitulate the assumptions we have made.

Theorem 3.13. *Let* $\mathcal{M} = (S, A, p(s'|s, a), r(s, a), \lambda)$ *be an infinite Markov decision problem where S is discrete, $0 \leq \lambda < 1$ and $r(s, a) \leq c$ for all $s \in S$ and $a \in A$. Then the value of $\mathcal{M}$ is v_λ^*, the unique solution of $\mathcal{B}v = v$.*

Proof. Note that $(V, \|\cdot\|_\infty)$ is a Banach space. Moreover $\mathcal{B} : V \to V$ is a contraction mapping according to Lemma 3.12. Hence we can use Banach's fixed-point theorem and obtain that $\mathcal{B}v = v$ has a unique solution. Now we know by Theorem 3.11 that the solution v^* of $\mathcal{B}v = v$ fulfils $v^* = v_\lambda^*$. $\qquad\square$

Next we prove that the Bellman equations can indeed be interpreted as optimality conditions for infinite Markov decision problems. We only have to summarize what we already know.

Theorem 3.14. *A policy $\pi^* \in \mathcal{D}^{RH}$ is optimal if and only if $v_\lambda^{\pi^*}$ is a solution of $\mathcal{B}v = v$.*

Proof. Suppose π^* is optimal. By definition we have $v_\lambda^{\pi^*} = v_\lambda^*$, therefore by Theorem 3.13 we know that $v_\lambda^{\pi^*}$ solves $\mathcal{B}v = v$. Now assume $v_\lambda^{\pi^*}$ solves the Bellman equations. Then by Theorem 3.11 we know $v_\lambda^{\pi^*} = v_\lambda^*$ hence π^* is optimal. $\qquad\square$

Now we want to convince ourselves that there exists an optimal policy of the form $\Delta = (d, d, \ldots)$, this kind of policy is called stationary. It uses the same decision rule at every decision epoch. That there are optimal policies which are stationary is also intuitively clear because transition probabilities and rewards are independent of time.

We are introducing some common terminology. We will start by assuming that the suprema in the Bellman equations are attained. In that case we introduce so called improving and conserving decision rules to show the existence of a stationary optimal policy.

Definition 3.15. *We call the decision rule $d_v \in D^{DM}$ v–improving if*

$$d_v \in \operatorname*{argmax}_{d \in D^{DM}} r_d + \lambda P_d v.$$

With D_v we denote the set of all v–improving decision rules.

This means a v–improving policy fulfils

$$r_{d_v} + \lambda P_{d_v} = \max_{d \in D^{DM}} r_d + \lambda P_d v$$

or, using the operator L_d in Definition 3.8 and the Bellman Operator we can rewrite this property as

$$L_{d_v} v = \mathcal{B} v.$$

Definition 3.16. *A decision rule $d^* \in D^{DM}$ is called conserving if it is v_λ^* improving.*

So conserving decision rules have the property that

$$\sup_{d \in D^{DM}} r_d + \lambda P_d v_\lambda^* = r_{d^*} + \lambda P_{d^*} v_\lambda^*.$$

Theorem 3.17. *Let S be discrete and suppose the supremum*

$$\sup_{d \in D^{DM}} r_d + \lambda P_d v$$

is attained for all $v \in V$. Then a policy of the form $\Delta := (d^*, d^*, \ldots)$, where $d^* \in D^{DM}$ is a conserving decision rule, is optimal.

Proof. We start by observing that a conserving decision rule exist. Note that $v_\lambda^* \in V$, hence $\sup r_d + \lambda P_d v_\lambda^*$ is attained by assumption. Now choose $d^* \in \operatorname{argmax} r_d + \lambda P_d v_\lambda^*$ then d^* is conserving by definition. By Theorem 3.13 and by the definition of conserving we have

$$v_\lambda^* = \mathcal{B}v_\lambda^* = \sup_{d \in D^{DM}} r_d + \lambda P_d v_\lambda^* = r_{d^*} + \lambda P_{d^*} v_\lambda^* = L_{d^*} v_\lambda^*.$$

Now by Theorem 3.7 we have $v_\lambda^\Delta = v_\lambda^*$, hence Δ is optimal. $\square$

So we finally have

$$\sup_{d \in D^{DM}} v_\lambda^\Delta = \sup_{\pi \in \mathcal{D}^{RH}} v_\lambda^\pi,$$

where $\Delta = (d, d, \ldots)$ is a stationary policy consisting of conserving decision rules. Now we have a look at another easy criteria ensuring the existence of conserving decision rules: Whenever there is an optimal policy there is also an optimal stationary policy.

Proposition 3.18. *Suppose there exists an optimal policy $\pi \in \mathcal{D}^{RH}$. Then there exists a deterministic stationary optimal policy Δ.*

Proof. Let $\pi^* = (d^*, \tilde{\pi})$ with[1] $d \in D^{RM}$ be an optimal policy. Then we have

$$\begin{aligned} v_\lambda^{\pi^*} = r_{d^*} + \lambda P_{d^*} v_\lambda^{\tilde{\pi}} &\leq r_{d^*} + \lambda P_{d^*} v_\lambda^{\pi^*} \\ &\leq \sup_{d \in D^{DM}} r_d + \lambda P_d v_\lambda^{\pi^*} \\ &= \mathcal{B}v_\lambda^{\pi^*} = v_\lambda^{\pi^*}, \end{aligned}$$

hence the inequalities are equalities and so d is a conserving decision rule. $\square$

---

[1] π^* is a randomized history dependent policy. Nevertheless d^* is randomized Markovian since the history up to decision epoch 1 equals the initial state. For the same reason stationary policies are always Markovian policies.

Note that in the proof of Theorem 3.17 the assumption that the supremum is attained was only needed to ensure the existence of a conserving decision rule. Now we have proved the existence of a conserving decision rule through the existence of an optimal policy.

In the case where the suprema are not attained we will show that there exists an ε–optimal stationary policy. We define them just like in the finite horizon case:

Definition 3.19. *Fix $\varepsilon > 0$. We call the policy π_ε^* ε–optimal if*

$$v_\lambda^{\pi_\varepsilon^*} \geq v_\lambda^* - \varepsilon \mathbf{1}.$$

Theorem 3.20. *Suppose S is discrete. Then for all $\varepsilon > 0$ there exists an ε–optimal deterministic stationary policy.*

Proof. Recall that $\sup_{d \in D^{DM}} r_d + \lambda P_d v_\lambda^* = \mathcal{B} v_\lambda^* = v_\lambda^*$. Now choose $d_\varepsilon \in D^{DM}$ so that

$$r_{d_\varepsilon} + \lambda P_{d_\varepsilon} v_\lambda^* \geq \sup_{d \in D^{DM}} r_d + \lambda P_d v_\lambda^* - (1 - \lambda)\varepsilon \mathbf{1} = v_\lambda^* - (1 - \lambda)\varepsilon \mathbf{1}.$$

This yields $r_{d_\varepsilon} \geq v_\lambda^* - \lambda P_{d_\varepsilon} v_\lambda^* \varepsilon \mathbf{1} - (1 - \lambda)\varepsilon \mathbf{1}$. Since $(\mathrm{id}_V - \lambda P_{d_\varepsilon})^{-1}$ is order preserving (Lemma 3.10) we have

$$(\mathrm{id}_V - \lambda P_{d_\varepsilon})^{-1} r_{d_\varepsilon}$$
$$\geq (\mathrm{id}_V - \lambda P_{d_\varepsilon})^{-1}(v_\lambda^* - \lambda P_{d_\varepsilon} v_\lambda^*) - (1 - \lambda)(\mathrm{id}_V - \lambda P_{d_\varepsilon})^{-1} \varepsilon \mathbf{1}.$$

Due to Theorem 3.7 we know that $v_\lambda^\Delta = (\mathrm{id}_V - \lambda P_{d_\varepsilon})^{-1} r_{d_\varepsilon}$ where $\Delta = (d_\varepsilon, d_\varepsilon, \ldots)$. We further have $(\mathrm{id}_V - \lambda P_{d_\varepsilon})^{-1}(v_\lambda^* - \lambda P_{d_\varepsilon} v_\lambda^*) = (\mathrm{id}_V - \lambda P_{d_\varepsilon})^{-1}(\mathrm{id}_V - \lambda P_{d_\varepsilon})v_\lambda^* = v_\lambda^*$ as well as

$$(\mathrm{id}_V - \lambda P_{d_\varepsilon})^{-1}\mathbf{1} = \sum_{t=1}^{\infty} \lambda^{t-1} P_{d_\varepsilon}^{t-1} \mathbf{1} = \frac{1}{1 - \lambda}\mathbf{1}.$$

So altogether we have $v_\lambda^\Delta \geq v_\lambda^* - \varepsilon \mathbf{1}$, hence $\Delta = (d_\varepsilon, d_\varepsilon, \ldots)$ is an ε–optimal policy. $\square$

Algorithm 4 Value Iteration Algoritm

Require: Infinite Markov Decision Problem $\mathcal{M}$, starting vector v_0, ε.

Ensure: An ε–optimal stationary policy $\Delta = (d_\varepsilon, d_\varepsilon, \ldots)$.

1: $v' \leftarrow v_0$.
2: **repeat**
3:    $v \leftarrow v'$
4:    **for all** $s \in S$ **do**
5:      $v'(s) = \max_{a \in A_s} r(s,a) + \lambda \sum_{j \in S} p(j|s,a) v(j)$.
6:    **end for**
7: **until** $\|v' - v\| \leq \frac{\varepsilon(1-\lambda)}{2\lambda}$
8: **for all** $s \in S$ **do**
9:    $d_\varepsilon \in \text{argmax}_{a \in A_s} r(s,a) + \lambda \sum_{j \in S} p(j|s,a) v'(j)$
10: **end for**

3.2 Value Iteration

Of course, our goal is to construct optimal policies. We will now take a closer look at the case where the suprema in the Bellman equation are attained and the state space S is finite. The Value Iteration algorithm then finds a stationary ε–optimal policy.

Definition 3.21. *Let $x_n \in V$ for all $n \in \mathbb{N}$ and $(x_n)_{n \in \mathbb{N}} \to x^*$ be a convergent sequence. The sequence converges at order $\alpha > 0$ if there exists a constant $K > 0$ for which we have*

$$\|x_{n+1} - x^*\| \leq K\|x_n - x^*\|^\alpha.$$

Let $(x_n)_{n \in \mathbb{N}}$ be a sequence converging at order α. The rate *of convergence is defined as the smallest K for which*

$$\|x_{n+1} - x^*\| \leq K\|x_n - x^*\|^\alpha.$$

Theorem 3.22. *Let $(v_n)_{n \in \mathbb{N}}$ be the sequence constructed by the Value Iteration algorithm. Then we have*

1. $v_n \to v_\lambda^*$ *for* $n \to \infty$.

2. *Convergence is linear with rate* λ.

3. *The policy* d_ε *constructed by the algorithm is* ε-*optimal.*

4. *For all* $n \in \mathbb{N}$ *we have*

$$\|v_n - v_\lambda^*\| \leq \frac{\lambda^n}{1 - \lambda} \|v_1 - v_0\|.$$

5. *For any* $d_n \in \operatorname{argmax} r_d + \lambda P_d v_n$ *we have*

$$\|v_\lambda^\Delta - v_\lambda^*\| \leq \frac{2\lambda^n}{1 - \lambda} \|v_1 - v_0\|,$$

where $\Delta = (d_n, d_n, \ldots)$.

6. *After* n_ε *steps the stopping criterion is definitely fulfilled with*

$$n_\varepsilon := \left\lceil \frac{\log\left(\frac{\varepsilon(1-\lambda)}{2\|v_1 - v_0\|}\right)}{\log \lambda} \right\rceil.$$

If S *and* A *are finite sets then the Value Iteration algorithm needs* $\mathcal{O}(n_\varepsilon |S|^2 |A|)$ *effort.*

Proof. 1. Observe that the value iteration algorithm is actually a fixed point iteration. Line 5 might be expressed in vector notation as

$$v' = \mathcal{B}v.$$

Therefore Banach's fixed–point theorem proves the claim.

2. We have $\mathcal{B}v_n = v_{n+1}$ and $\mathcal{B}v_\lambda^* = v_\lambda^*$, therefore $\|v_{n+1} - v_\lambda^*\| = \|\mathcal{B}v_n - \mathcal{B}v_\lambda^*\| \leq \lambda\|v_n - v_\lambda^*\|$ since $\mathcal{B} : V \to V$ is a contraction mapping and hence convergence is linear. Now choosing $v_0 := v_\lambda^* + 1$ yields $v_1 = \max_{d \in D^{DM}} r_d + \lambda P_d v_0 = \max_{d \in D^{DM}} r_d + \lambda P_d(v_\lambda^* + 1) = v_\lambda^* + \lambda 1$. Now we have $v_1 - v_\lambda^* = \lambda 1 = \lambda(v_0 - v_\lambda^*)$ hence the convergence rate is λ.

3. We have $\|v_\lambda^\Delta - v_\lambda^*\| \leq \|v_\lambda^\Delta - v_{n+1}\| + \|v_{n+1} - v_\lambda^*\|$, where $\Delta = (d_\varepsilon, d_\varepsilon, \ldots)$. Observe that we further have $L_{d_\varepsilon} v_\lambda^\Delta = v_\lambda^\Delta$ as well as $L_{d_\varepsilon} v_{n+1} = \mathcal{B} v_{n+1}$. Now we calculate

$$
\begin{aligned}
\|v_\lambda^\Delta - v_{n+1}\| &= \|L_{d_\varepsilon} v_\lambda^\Delta - v_{n+1}\| \\
&\leq \|L_{d_\varepsilon} v_\lambda^\Delta - \mathcal{B} v_{n+1}\| + \|\mathcal{B} v_{n+1} - v_{n+1}\| \\
&= \|L_{d_\varepsilon} v_\lambda^\Delta - L_{d_\varepsilon} v_{n+1}\| + \|\mathcal{B} v_{n+1} - \mathcal{B} v_n\| \\
&\leq \lambda \|v_\lambda^\Delta - v_{n+1}\| + \lambda \|v_{n+1} - v_n\|.
\end{aligned}
$$

This yields $\|v_\lambda^\Delta - v_{n+1}\| \leq \frac{\lambda}{1-\lambda} \|v_{n+1} - v_n\| < \frac{\varepsilon}{2}$. Now we need a similar argument to estimate

$$
\begin{aligned}
\|v_\lambda^* - v_{n+1}\| &= \|\mathcal{B} v_\lambda^* - v_{n+1}\| \\
&\leq \|\mathcal{B} v_\lambda^* - \mathcal{B} v_{n+1}\| + \|\mathcal{B} v_{n+1} - v_{n+1}\| \\
&= \|\mathcal{B} v_\lambda^* - \mathcal{B} v_{n+1}\| + \|\mathcal{B} v_{n+1} - \mathcal{B} v_n\| \\
&\leq \lambda \|v_\lambda^* - v_{n+1}\| + \lambda \|v_{n+1} - v_n\|.
\end{aligned}
$$

Again we have $\|v_\lambda^* - v_{n+1}\| \leq \frac{\lambda}{1-\lambda} \|v_{n+1} - v_n\| < \frac{\varepsilon}{2}$. Therefore, putting the two estimates together we have $\|v_\lambda^\Delta - v_\lambda^*\| \leq \frac{2\lambda}{1-\lambda} \|v_{n+1} - v_n\| < \varepsilon$ which yields $v_\lambda^* \leq v_\lambda^\Delta + \mathbf{1}\varepsilon$, hence Δ is an ε–optimal policy.

4. Using (3) and the contraction property proves the claim:

$$
\begin{aligned}
(1 - \lambda)\|v_n - v_\lambda^*\| \leq \lambda \|v_n - v_{n-1}\| &= \lambda \|\mathcal{B} v_{n-1} - \mathcal{B} v_{n-2}\| \\
&\leq \lambda^2 \|v_{n-1} - v_{n-2}\| \\
&\leq \lambda^n \|v_1 - v_0\|.
\end{aligned}
$$

5. Again using (3) and the contraction property proves the claim:

$$
\|v_\lambda^\Delta - v_\lambda^*\| \leq \frac{2\lambda}{1-\lambda} \|v_n - v_{n-1}\| \leq \cdots \leq \frac{2\lambda^n}{1-\lambda} \|v_1 - v_0\|.
$$

6. The statement in (5) yields a worst case estimate of how often the repeat–until loop has to be executed to definitely fulfil the stopping criterion. Solving the equation

$$
\frac{2\lambda^n}{1-\lambda} \|v_1 - v_0\| \leq \varepsilon
$$

for n gives the desired quantity n_ε. Now like in the finite case a maximization has to be done. If A is finite we have a worst case estimate: we simply try out every possible value $a \in A$ for a fixed $s \in S$. To be able to compare to other actions we need to perform a function evaluation which itself is in $\mathcal{O}(|S|)$ due to the summation over all $s \in S$. Hence we have a total effort of $\mathcal{O}(n_\varepsilon |S|^2 |A|)$.

$\square$

3.3 Policy Iteration

The second important algorithm for infinite Markov decision problems is called Policy Iteration. Here we start with an arbitrary policy and in every iteration step this policy is improved.

Algorithm 5 Policy Iteration Algorithm

Require: Infinite Markov Decision Problem $\mathcal{M}$, starting decision rule $d_0 \in D^{DM}$.

Ensure: Optimal decision rule.

1: $d' \leftarrow d_0$.
2: **repeat**
3:    $d \leftarrow d'$.
4:    $v = (\mathrm{Id}_V - \lambda P_d)^{-1} r_d$.
5:    Choose $d' \in \mathrm{argmax}_{d \in D^{DM}} r_d + \lambda P_d v$
6: **until** $d' = d$

To analyse policy iteration we introduce the operator $\mathcal{P} : V \to V$ where

$$\mathcal{P}v := \max_{d \in D^{DM}} r_d + (\lambda P_d - \mathrm{id}_V)v.$$

Note that we now have $\mathcal{P}v = \mathcal{B}v - v$, hence the Bellman equations can be expressed by the equation $\mathcal{P}v = 0$.

Proposition 3.23. *For $u, v \in V$ and any v–improving decision rule d_v we have*

$$\mathcal{P}u \geq \mathcal{P}v + (\lambda P_{d_v} - id_V)(u - v). \tag{3.10}$$

Proof. Recall that d_v is v–improving if $d_v \in \text{argmax}_{d \in D^{DM}} r_d + \lambda P_d v = \text{argmax}_{d \in D^{DM}} r_d + (\lambda P_d v - \text{id}_V)v$. Now by definition we have

$$\mathcal{P}u \geq r_{d_v} + (\lambda P_{d_v} - \text{id}_V)u,$$

as well as

$$\mathcal{P}v = r_{d_v} + (\lambda P_{d_v} - \text{id}_V)v.$$

Now substracting the equality from the inequality gives $\mathcal{P}u \geq \mathcal{P}v + (\lambda P_{d_v} - \text{id}_V)(u - v)$.   □

Now we take a closer look at the sequence $(v_n)_{n \in \mathbb{N}}$ constructed by the Policy Iteration algorithm.

Proposition 3.24. *Let v_n be successively generated by Algorithm 5. Then we have*

1. $v_{n+1} \geq v_n$.

2. For any $d \in D_{v_n}$ we have

$$v_{n+1} = v_n - (\lambda P_d - \text{id}_V)^{-1}\mathcal{P}v_n.$$

Proof. Let d_n and v_n be constructed by the Policy Iteration.

1. Observe that we have $(\text{id}_V - \lambda P_{d_n})v_n = r_{d_n}$ which means $v_n = \lambda P_{d_n} v_n + r_{d_n}$. Then it follows that $r_{d_{n+1}} + \lambda P_{d_{n+1}} v_n \geq r_{d_n} + \lambda P_{d_n} v_n = v_n$, and hence

$$r_{d_{n+1}} \geq (\text{id}_V - \lambda P_{d_{n+1}})v_n.$$

Now we use Lemma 3.10 with $u = r_{d_{n+1}}$ and $v = (\text{id}_V - \lambda P_{d_{n+1}})v_n$ and obtain

$$v_{n+1} = (\text{id}_V - \lambda P_{d_{n+1}})^{-1}r_{d_{n+1}} \geq v_n.$$

2. Note that $d := d_{n+1}$ is a v_n improving decision rule (we already needed this fact for the very first estimate in the proof of (i)). We have

$$
\begin{aligned}
v_{n+1} &= (\mathrm{id}_V - \lambda P_{d_{v_{n+1}}})^{-1} r_{d_{n+1}} - v_n + v_n \\
&= (\mathrm{id}_V - \lambda P_d)^{-1} (r_d - (\mathrm{id}_V - \lambda P_d)v_n) + v_n \\
&= (\mathrm{id}_V - \lambda P_d)^{-1} (r_d + (\lambda P_d - \mathrm{id}_V)v_n) + v_n \\
&= (\mathrm{id}_V - \lambda P_d)^{-1} \mathcal{P} v_n + v_n \\
&= v_n - (\lambda P_d - \mathrm{id}_V)^{-1} \mathcal{P} v_n.
\end{aligned}
$$

$\square$

Now Proposition 3.24 gives us a direct formula for v_{n+1} in terms of v_n if we assume that we already know a v_n–improving decision rule $d \in D_{v_n}$. This update formula can be seen as a generalization of Newtons method $x_{n+1} = x_n - f'(x)^{-1} f(x_n)$ for finding zeros of a function f. If f is convex and has a zero $\hat{x}$ as well as $f(x_0) > 0$ then $(x_n)_{n \in \mathbb{N}}$ is a monotonically decreasing sequence with $x_n \to \hat{x}$. Now (3.10) somehow looks similar to $f(x) \geq f(y) + f'(y)(x - y)$, a property that characterizes convex functions if fulfilled for every x, y in the domain of f. We are interested in solving $\mathcal{P}v = 0$ since this equation is equivalent to the Bellman equation $\mathcal{B}v = v$. This analogy gives rise to define $\mathcal{N} : V \to V$,

$$
\mathcal{N}v := v + (\mathrm{id}_V - \lambda P_{d_v})^{-1} \mathcal{P}v.
$$

We further define $V_{\mathcal{P}} := \{v \in V : \mathcal{P}v \geq 0\}$. Note that if $v \in V_{\mathcal{P}}$ we have $\mathcal{P}v = \mathcal{B}v - v \geq 0$ and by the proof of Theorem 3.11 $v \leq v_\lambda^*$. So whenever $v \in V_{\mathcal{P}}$ then v is a lower bound of v_λ^*.

Before we are able to prove convergence of the policy iteration we need the following lemma.

Lemma 3.25. *Let $v \in V_{\mathcal{P}}$, choose $d_v \in D_v$ and assume $v \geq u$. Then we have*

1. $\mathcal{N}v \geq \mathcal{B}u$.

2. $\mathcal{N}v \in V_{\mathcal{P}}$.

Proof. 1. First note that if $v \geq u$ there exists a vector $\epsilon \geq 0$ with $v = u + \epsilon$. Now $\mathcal{B}v = \mathcal{B}(u + \epsilon) = \sup_{d \in DDM} r_d + P_d u + P_d \epsilon \geq \sup_{d \in DDM} r_d + P_d u = \mathcal{B}u$ because $P_d \epsilon \geq 0$. Since $v \in V_{\mathcal{P}}$ we have $\mathcal{P}v \geq 0$ and therefore by Lemma 3.10, where we set $x = \mathcal{P}v$, we see that

$$\mathcal{N}v = v + (\mathrm{id}_V - \lambda P_d)^{-1} \mathcal{P}v \geq v + \mathcal{P}v = \mathcal{B}v \geq \mathcal{B}u.$$

2. We are using Proposition 3.23 with $u = \mathcal{N}(v)$ to obtain

$$\mathcal{P}(\mathcal{N}v) \geq \mathcal{P}v + (\lambda P_{d_v} - \mathrm{id}_V)(\mathcal{N}v - v) = \mathcal{P}v - \mathcal{P}v = 0.$$

□

Now we are ready to provide a convergence result for the Policy Iteration.

Theorem 3.26. *The sequence $(v_n)_{n \in \mathbb{N}}$ generated by the Policy Iteration algorithm converges monotonically to v_λ^*.*

Proof. We define $u_k := \mathcal{B}^k v_0$. By Theorem 3.22 we already know that $\|u_k - v_\lambda^*\| \to 0$ if $k \to \infty$. We now show inductively, that

$$u_k \leq v_k \leq v_\lambda^* \text{ and } v_k \in V_{\mathcal{P}}.$$

If this is satisfied $\|v_k - v_\lambda^*\| \to 0$ which yields the claim.

Now let $k = 0$. We have $\mathcal{P}v_0 \geq r_{d_0} + (\lambda P_{d_0} - \mathrm{id}_V)v_0 = r_{d_0} + (\lambda P_{d_0} - \mathrm{id}_V)(\mathrm{id}_V - \lambda P_{d_0})^{-1} r_{d_0} = 0$, hence $v_0 \in V_{\mathcal{P}}$. This furthermore yields $\mathcal{P}v_0 = \mathcal{B}v_0 - v_0 \geq 0$ and therefore $\mathcal{B}v_0 \geq v_0$ which implies, using the proof of Theorem 3.11, $v_0 \leq v_\lambda^*$. Furthermore we have $u_0 = \mathcal{B}^0 v_0 = v_0$, therefore $u_0 \leq v_0$.

Now lets assume that the result holds for all $k \leq n$. The iterates of the Policy Iteration fulfil $v_{n+1} = \mathcal{N}v$. By Lemma 3.25 and by the induction hypothesis we have that $v_{n+1} \in V_{\mathcal{P}}$ as well as $v_{n+1} = \mathcal{N}v_n \geq \mathcal{B}u_n = u_{n+1}$. Finally, we also have $\mathcal{P}v_{n+1} = \mathcal{P}(\mathcal{N}(v_n)) \geq 0$ and analogously to the induction basis we have $\mathcal{B}v_{n+1} \geq v_{n+1}$ and therefore again $v_{n+1} \leq v_\lambda^*$.

Monotonicity of $(v_n)_{n \in \mathbb{N}}$ follows from Proposition 3.24. □

Now the problem is that we have no guarantee that the stopping criterion in Algorithm 5 is fulfilled in finite time. Therefore, we again have to be content with ε–optimal policies. To construct ε–optimal policies we just use the stopping criterion of the Value Iteration algorithm. So Algorithm 5 has to be slightly adapted. We now have to store not only the current v_n but also v_{n-1} to be able to compare these two values.

Algorithm 6 ε–Optimal Policy Iteration Algorithm

Require: Infinite Markov Decision Problem $\mathcal{M}$, starting decision rule $d_0 \in D^{DM}$, ε.
Ensure: ε–optimal policy.
1: $d' \leftarrow d_0$
2: $v' = (\text{Id}_V - \lambda P_d)^{-1} r_d$
3: **repeat**
4: $\quad v \leftarrow v'$
5: $\quad v' = (\text{Id}_V - \lambda P_d)^{-1} r_d$
6: $\quad$ Choose $d' \in \text{argmax}_{d \in D^{DM}} r_d + \lambda P_d v'$
7: $\quad d \leftarrow d'$
8: **until** $\|v' - v\| \leq \varepsilon \frac{1-\lambda}{2\lambda}$

The proof for ε–optimality is the same as we had in Theorem 3.22 since the stopping criterion is the same.

Now we want to take a closer look at the convergence rate of the Policy Iteration algorithm. We already know that convergence is at least linear since convergence of Value Iteration is linear and the iterates produced by Value Iteration are always smaller or equal to the iterates produced by Policy Iteration (and both Value and Policy Iteration produce iterates bounded from above by v_λ^*). Under certain conditions we have quadratic convergence.

Theorem 3.27. *Let $(v_n)_{n \in \mathbb{N}}$ be the sequence generated by the Policy Iteration algorithm and $d_n \in D_{v_n}$. Suppose further that*

$$\|P_{d_n} - P_{d^*}\| \leq c \|v_n - v_\lambda^*\|,$$

where $0 < c < \infty$ and define $P_{d^*} := P_{v_\lambda^*}$.
    Then we have

$$\|v_{n+1} - v_\lambda^*\| \leq \frac{c\lambda}{1 - \lambda}\|v_n - v_\lambda^*\|^2.$$

Proof. Using Proposition 3.23 we have

$$\mathcal{P}v_n \geq \mathcal{P}v_\lambda^* + (\lambda P_{d^*} - \mathrm{id}_V)(v_n - v_\lambda^*) \geq (\lambda P_{d^*} - \mathrm{id}_V)(v_n - v_\lambda^*).$$

Now this yields with Lemma 3.10

$$(\mathrm{id}_V - \lambda P_{d_n})^{-1}\mathcal{P}v_n \geq (\mathrm{id}_V - \lambda P_{d_n})^{-1}(\lambda P_{d^*} - \mathrm{id}_V)(v_n - v_\lambda^*)$$

and hence

$$(\lambda P_{d_n} - \mathrm{id}_V)^{-1}\mathcal{P}v_n \leq (\lambda P_{d_n} - \mathrm{id}_V)^{-1}(\lambda P_{d^*} - \mathrm{id}_V)(v_n - v_\lambda^*)$$

because $-(\lambda P_{d_n} - \mathrm{id}_V)^{-1} = (\mathrm{id}_V - \lambda P_{d_n})^{-1}$. Using this inequality we can estimate

$$
\begin{aligned}
v_\lambda^* &- v_{n+1} \\
&= v_\lambda^* - \mathcal{N}v_n \\
&= v_\lambda^* - v_n + (\lambda P_{d_n} - \mathrm{id}_V)^{-1}\mathcal{P}v_n \\
&\leq (\lambda P_{d_n} - \mathrm{id}_V)^{-1}(\lambda P_{d_n} - \mathrm{id}_V)(v_\lambda^* - v_n) \\
&\quad + (\lambda P_{d_n} - \mathrm{id}_V)^{-1}(\lambda P_{d^*} - \mathrm{id}_V)(v_n - v_\lambda^*) \\
&= (\lambda P_{d_n} - \mathrm{id}_V)^{-1}(\lambda P_{d_n} - \lambda P_{d^*})(v_\lambda^* - v_n).
\end{aligned}
$$

Now this yields

$$\|v_\lambda^* - v_{n+1}\| \leq \|(\lambda P_{d_n} - \mathrm{id}_V)^{-1}\|\|\lambda P_{d_n} - \lambda P_{d^*}\|\|v_\lambda^* - v_n\|$$
$$\leq \frac{1}{1 - \lambda}\lambda c\|v_\lambda^* - v_n\|^2,$$

because $\lambda\|P_{d_n} - P_{d^*}\| \leq c\lambda\|v_\lambda^* - v_n\|$ by assumption and

$$\|(\lambda P_{d_n} - \mathrm{id}_V)^{-1}\| = \|(\mathrm{id}_V - \lambda P_{d_n})^{-1}\|$$

$$= \sup_{\|v\| \leq 1} \|(\mathrm{id}_V - \lambda P_{d_n})^{-1} v\|$$

$$= \sup_{\|v\| \leq 1} \|\sum_{t=1}^{\infty} \lambda^{t-1} P_{d_n}^{t-1} v\|$$

$$\leq \|\sum_{t=1}^{\infty} \lambda^{t-1} P_{d_n}^{t-1} \mathbf{1}\|$$

$$\leq \|\sum_{t=1}^{\infty} \lambda^{t-1} \mathbf{1}\|$$

$$= \frac{1}{1 - \lambda}.$$

$\square$

4 Markov Decision Problems and Clinical Trials

In this chapter we want to investigate the applicability of the MDP-theory to clinical trials. We want to do that in a response adaptive way so that the future trial members already benefit from the previous ones. The goal is to identify the better treatment and keep the number of trial members treated with the inferior therapy small. In [BE95] and [HS91] we find an approach using Bandit models which are similar to Markov decision problems. In [Pre09] some ethical cost models are introduced. In [Put94] and in [BR11] we find an overview how Bandit models can be treated as Markov decision problems. Both suggest that possible applications are clinical trials. In the following we give a detailed description how a possible implementation of a Markov decision problem for clinical trials looks like and provide numerical results.

We want to compare two medical treatments T_1 and T_2 with two unknown success probabilities p_1 and p_2. In the trial we have altogether M patients and we want to sequentially allocate them to one of the two treatments.

We assume dichotomous outcomes which can be observed immediately: after every allocation we observe either a success $\mathfrak{s}$ or a fail $\mathfrak{f}$. Allocation of the $(n+1)$-st person should be based on the knowledge we gained so far. The knowledge k_n at time point n is a four-tupel,

$$k_n := (s_1, f_1, s_2, f_2),$$

where s_1 is the number of patients allocated to treatment T_1 and a success $\mathfrak{s}$ is observed whereas f_1 is the number of patients allocated to treatment T_1 where a negative effect $\mathfrak{f}$ could be measured. The quantities s_2 and f_2 for treatment T_2 are defined analogously. We define the set of all possible knowledge

$$\mathcal{K}_M^4 := \{(s_1, f_1, s_2, f_2) : s_i, f_i \in \mathbb{N}, s_1 + f_1 + s_2 + f_2 \leq M\}.$$

$\mathcal{K}_M^4$ is the integer valued four dimensional simplex with edge length M. After we have allocated the $(n+1)$-st person we update our knowledge correspondingly, e.g., if we have allocated the person to treatment T_2 and a positive effect could be observed we have $k_{n+1} = (s_1, f_1, s_2 + 1, f_2)$ and if a negative effect could be observed we have $k_{n+1} = (s_1, f_1, s_2, f_2 + 1)$. We normally have a starting knowledge of $k_0 = (0, 0, 0, 0)$.

The goal is to find a good allocation policy which takes into account the following two things:

1. In the end the treatment with the higher efficacy should be identified.

2. We want to have as few patient losses as possible. This means that the majority of trial members should receive the better of the two treatments.

The first requirement is the one you expect from a clinical trial. Randomized assignment of patients to treatments is the standard in clinical practice. In the end of the trial some statistical hypothesis tests at a high level of statistical significance are made for establishing the effectiveness of medical treatments. Nevertheless, since allocation is randomized there will always be many patients assigned to the inferior treatment which leads to ethical problems. The second requirement counteracts this issue and has the advantage that more volunteers are willing to participate in a clinical trial since they know that they will be treated – in some sense – to the best of one's knowledge. Here *in some sense* is determined by the ingredients of the underlying Markov decision problem. It depends on the set of actions, the set of states, the transition probabilities, and the rewards.

We start with a short overview of ethical cost models based on [Pre09] where the allocation problem is treated with the help of Bandit models. Let c_1 and c_2 be the costs for treating a patient with T_1 and T_2 respectively. They should only depend on the unknown success probabilities p_1 and p_2 what leads to an ethical choice. *Expected failures* assigns the costs for treating a patient in the following way:

$$c_1(p_1, p_2) = 1 - p_1, \quad c_2(p_1, p_2) = 1 - p_2,$$

hence we have high costs if we choose a treatment with a small success probability. Then there is a cost model called *expected successes lost*, this assigns the difference of the success probabilities only if we choose the worse of the two treatments according to

$$c_1(p_1, p_2) = \begin{cases} 0 & \text{if } p_1 \geq p_2 \\ p_2 - p_1 & \text{if } p_1 < p_2 \end{cases},$$

$$c_2(p_1, p_2) = \begin{cases} 0 & \text{if } p_2 \geq p_1 \\ p_1 - p_2 & \text{if } p_2 < p_1. \end{cases}$$

Let D_1 and D_2 denote the costs of treatment T_1 and T_2. Including the costs for a treatment leads to *Dollar cost of failure*,

$$c_1(p_1, p_2) = (1 - p_1)D_1, \quad c_2(p_1, p_2) = (1 - p_2)D_1. \tag{4.1}$$

This cost function is kind of problematic since it could in principle be that the worse treatment still has lower costs due to the lower treatment cost. Better is the following choice

$$c_1(p_1, p_2) = \begin{cases} 0 & \text{if } p_1 \geq p_2 \\ (p_2 - p_1)D_1 & \text{if } p_1 < p_2 \end{cases},$$

$$c_2(p_1, p_2) = \begin{cases} 0 & \text{if } p_2 \geq p_1 \\ (p_1 - p_2)D_2 & \text{if } p_2 < p_1, \end{cases}$$

called *expected cost of treatment of lost successes*. This model only assigns costs if the worse of the two treatments is used.

Markov decision problems somehow solve the "exploration versus exploitation" dilemma. If you want to assign the next patient to a treatment and you know that, lets say T_1, has a high success probability then you can have a high immediate reward—you exploit the information you already have. Sometimes it might be better if you assign the treatment from which you do not have a clear impression how good it is. Then you explore some information by assigning T_2 hoping that it has an even higher success probability.

In the next section we want to describe how we choose the state space, the set of actions, the transition probabilities, and the rewards.

4.1 Finite Horizon Markov Decision Problem Formulation

We have two actions in every decision epoch, namely allocating the current patient to one of the treatments, hence $A := \{T_1, T_2\}$. Furthermore, we set the horizon length to $T := M + 1$, where M is the number of participants in the trial. This is because we want to allocate all of the M patients and at the last time point $M + 1$ by definition we get a final reward but no action has to be chosen.

Finding an appropriate state space is more tricky. We will encode our belief about the success probabilities in the state space. We consider the unknown probabilities p_1 and p_2 themselves to be random variables and express our knowledge about these success probabilities through probability densities. Given two densities f_1 and f_2 for p_1 and p_2 the corresponding Bayesian estimators $\widehat{p_1}$ and $\widehat{p_2}$ are given by

$$\widehat{p_1} = E_{f_1}(X) = \int_0^1 x f_1(x)\, \mathrm{d}x,$$

$$\widehat{p_2} = E_{f_2}(X) = \int_0^1 x f_2(x)\, \mathrm{d}x.$$

With $\mathscr{D}$ we denote the set of all probability densities with support on $[0, 1]$ and define the state space $S := \mathscr{D} \times \mathscr{D}$. Hence, an element of the state space is a pair of probability densities, $s = (f_1, f_2)$ reflecting our belief of the unknown success probabilities. After every allocation of one patient to one treatment the state changes.

Now we want to define transition probabilities. The next state $s' = (f_1', f_2')$ is allowed to depend on the current state $s = (f_1, f_2)$ and the chosen action $a \in A = \{T_1, T_2\}$ which results in an observation, namely if the treatment was successful or not. We assume that this observation can be made immediately after we medicate the patient. If we decide to use treatment T_1 we certainly do not gain additional information about the second treatment, hence $s' = (f_1', f_2)$ so we do not change our belief about the treatment which is currently not used. If we choose treatment T_2 the same is true, meaning that the density for the first treatment T_1 will stay the same, but the density for the second treatment is allowed to change.

Two questions arise: How should the part of the state space which is allowed to change be updated and what is the right probability therefore.

Assume we have decided to use treatment T_1. After this allocation we observe either $\mathfrak{s}$ or $\mathfrak{f}$. Based on this binary observation we want to update f_1. We define the binary random variable $Y : \Omega \to \{\mathfrak{s}, \mathfrak{f}\}$ where $P(Y = \mathfrak{s}) := \widehat{p_1}$ and $P(Y = \mathfrak{f}) = 1 - \widehat{p_1}$ and define the formula for the posterior density given a binary observation Y and a prior density f_1,

$$f_1(p_1 = x | Y = y) := \frac{P(Y = y | p_1 = x) f_1(x)}{P(Y = y)}$$

We derive the posterior densities according to the definition,

$$f_1'(x) = f_1(x|\mathfrak{s}) = \frac{x f_1(x)}{\widehat{p_1}},$$

if a success $\mathfrak{s}$ is observed and

$$f_1'(x) = f_1(x|\mathfrak{f}) = \frac{(1 - x) f_1(x)}{1 - \widehat{p_1}}$$

if a fail $\mathfrak{f}$ is observed.

Remark 4.1. Indeed f_1' is a probability density on $\Omega = [0,1]$. Consider for example

$$\int_0^1 \frac{(1-x)f_1(x)}{1-\widehat{p_1}}\, \mathrm{d}x = \frac{1}{1-\widehat{p_1}}\left(\int_0^1 f_1(x)\,\mathrm{d}x - \int_0^1 xf_1(x)\,\mathrm{d}x\right)$$

$$= \frac{1}{1-\widehat{p_1}}(1-\widehat{p_1}) = 1,$$

because f_1 is assumed to be a probability density on $[0,1]$. One easily calculates that also $\int_0^1 f_1'(x|\mathbf{s})\,\mathrm{d}x = 1$. Moreover $f_1' \geq 0$ on $[0,1]$, so f_1' is a probability density. Analogously we have that f_2' is a probability density.

Now, if we choose treatment T_1, to a given state $s = (f_1, f_2)$ we have $s' = (f_1', f_2)$ where we expect $f_1' = f_1(x|\mathbf{s})$ with probability $\widehat{p_1}$ and $f_1' = f_2(x|\mathbf{f})$ with probability $1 - \widehat{p_1}$. So to a given density f_1 we only allow two possible succeeding densities. Then the updated density f_1' will also change the estimator for p_1 since we then assume a new density for it.

These considerations lead to the transition probabilities

$$p(s' = (f_1', f_2')|s = (f_1, f_2), a = T_1)$$

$$= \begin{cases} \widehat{p_1} & \text{if } s' = \left(\frac{xf_1(x)}{\widehat{p_1}}, f_2\right) \\ 1 - \widehat{p_1} & \text{if } s' = \left(\frac{(1-x)f_1(x)}{1-\widehat{p_1}}, f_2\right) \\ 0 & \text{otherwise,} \end{cases} \qquad (4.2)$$

and analogously for treatment T_2 we have

$$p(s' = (f_1', f_2')|s = (f_1, f_2), a = T_2)$$

$$= \begin{cases} \widehat{p_2} & \text{if } s' = \left(f_1, \frac{xf_2(x)}{\widehat{p_2}}\right) \\ 1 - \widehat{p_2} & \text{if } s' = \left(f_1, \frac{(1-x)f_2(x)}{1-\widehat{p_2}}\right) \\ 0 & \text{otherwise.} \end{cases} \qquad (4.3)$$

Definition 4.2 (Beta function and beta distribution). *The function*

$$B(a,b) := \int_0^1 x^{a-1}(1-x)^{b-1}\,\mathrm{d}x$$

is called beta function.

The probability density function $f_{a,b}$ of the (a, b)–beta distribution is given by

$$f_{a,b}(x) := \begin{cases} \frac{1}{B(a,b)} x^{a-1}(1-x)^{b-1} & \text{if } 0 \le x \le 1 \\ 0 & \text{otherwise.} \end{cases}$$

Lemma 4.3. *Fix $i \in \{1, 2\}$ and let f_i be a density of the (a, b)–beta distribution. Then we have that*

1. $f_i(x|\mathfrak{s})$ *is a density of the $(a+1, b)$–beta distribution.*

2. $f_i(x|\mathfrak{f})$ *is a density of the $(a, b+1)$–beta distribution.*

3. *If we observe s successes and f fails in $s + f$ trials then the corresponding $(s + f)$–times updated density is the density of an $(a + s, b + f)$–beta distribution.*

4. *If X is (a, b)–beta distributed we have $E(X) = \frac{a}{a+b}$.*

Proof. 1. We have

$$f_i(x|\mathfrak{s}) = \frac{x f_i(x)}{\widehat{p}_i} = \frac{x^a(1-x)^{b-1}}{B(a,b) \int_0^1 \frac{x}{B(a,b)} x^{a-1}(1-x)^{b-1} \, \mathrm{d}x}$$

$$= \frac{x^a(1-x)^{b-1}}{B(a+1, b)}$$

$$= f_{a+1,b}(x).$$

2. First of all we have, by definition of $B(a, b)$ that

$$\int_0^1 \frac{x^{a-1}(1-x)^{b-1}}{B(a,b)} \, \mathrm{d}x = 1,$$

and hence

$$1 - \widehat{p}_i = \int_0^1 \frac{x^{a+1}(1-x)^{b-1}}{B(a,b)} \, \mathrm{d}x - \int_0^1 \frac{x}{B(a,b)} x^{a-1}(1-x)^{b-1} \, \mathrm{d}x$$

$$= \int_0^1 (1-x) \frac{x^{a-1}(1-x)^{b-1}}{B(a,b)} \, \mathrm{d}x$$

$$= \frac{B(a,b+1)}{B(a,b)}.$$

Finally we have

$$f_{T_i}(x|\mathfrak{f}) = \frac{(1-x)f_{T_i}(x)}{1-\widehat{p}_i} = \frac{(1-x)x^{a-1}(1-x)^{b-1}}{B(a,b)(1-\widehat{p}_i)}$$

$$= \frac{1}{B(a,b+1)} x^{a-1}(1-x)^b$$

$$= f_{a,b+1}(x).$$

3. Follows from (i) and (ii) if applied iteratively.

4. We have to use the identity

$$B(a,b) = \frac{\Gamma(a)\Gamma(b)}{\Gamma(a+b)},$$

where Γ denotes the Gamma function, as well as the functional equation $\Gamma(a+1) = a\Gamma(a)$. Then we easily calculate

$$E(X) = \frac{1}{B(a,b)} \int_0^1 x^a(1-x)^{b-1} \, \mathrm{d}x = \frac{\Gamma(a+1)\Gamma(b)}{B(a,b)\Gamma(a+b+1)}$$

$$= \frac{a\Gamma(a)\Gamma(b)}{(a+b)B(a,b)\Gamma(a+b)}$$

$$= \frac{a}{a+b}.$$

$\square$

This now yields, from a computational perspective, to an important simplification. If we start with a beta distributed prior, i.e., with a density of a beta distribution all the further densities will also be densities of beta distributions if updated according to (4.2) and (4.3).

This means that we do not have to use $S = \mathscr{D} \times \mathscr{D}$, which would be computationally infeasible. Instead we use our knowledge space

$$S := \mathscr{K}_M^4$$

defined at the beginning of this chapter. Given the initial parameters (a_1, b_1) and (a_2, b_2), needed for the prior beta distributions, every element of $\mathscr{K}_M^4$ now defines a pair of beta distributions through $\varphi : \mathscr{K}_M^4 \to \mathscr{D} \times \mathscr{D}$ given by

$$\varphi((s_1, f_1, s_2, f_2)) = (f_{a_1+s_1, b_1+f_1}, f_{a_2+s_2, b_2+f_2}).$$

Conversely, if we start with a pair of (a_i, b_i)–beta distributions, where $i = \{1, 2\}$ every possible[1] pair of probability densities defines an element of $\mathscr{K}_M^4$. Since we know by Lemma 4.3 that the possible probability density pairs are of the form $(f_{a_i+l, b_i+(n-l)})_{i=1,2}$ the mapping φ is invertible if $\mathscr{D} \times \mathscr{D}$ is restricted to the possible probability densities. Then we have

$$\varphi^{-1}(f_{\alpha,\beta}, f_{\gamma,\delta}) = (\alpha - a_1, \beta - b_1, \gamma - a_2, \delta - b_2).$$

Now via this bijection we define the transition probabilities for $S = \mathscr{K}_M^4$,

$$\begin{aligned}
p((s_1', f_1', s_2', f_2')|(s_1, f_1, s_2, f_2), a) \\
:= p(\varphi(s_1', f_1', s_2', f_2')|\varphi(s_1, f_1, s_2, f_2), a),
\end{aligned}$$

what leads to

$$p((s_1', f_1', s_2', f_2')|(s_1, f_1, s_2, f_2), a = T_1)$$
$$= \begin{cases} \widehat{p_1} & \text{if } s_1' = s_1 + 1 \wedge f_1' = f_1 \wedge s_2' = s_2 \wedge f_2' = f_2 \\ 1 - \widehat{p_1} & \text{if } s_1' = s_1 \wedge f_1' = f_1 + 1 \wedge s_2' = s_2 \wedge f_2' = f_2 \\ 0 & \text{else} \end{cases} \quad (4.4)$$

---

[1] Possible means here that the probability that a pair of densities occurs in a state at an arbitrary time is not zero.

and

$$p((s_1', f_1', s_2', f_2')|(s_1, f_1, s_2, f_2), a = T_2)$$

$$= \begin{cases} \widehat{p_2} & \text{if } s_1' = s_1 \wedge f_1' = f_1 \wedge s_2' = s_2 + 1 \wedge f_2' = f_2 \\ 1 - \widehat{p_2} & \text{if } s_1' = s_1 \wedge f_1' = f_1 \wedge s_2' = s_2 \wedge f_2' = f_2 + 1 \quad (4.5) \\ 0 & \text{else} \end{cases}$$

If we don't have a clue which of the treatments is better in advance, what is usually the case, we can use the uninformative prior $a_1 = b_1 = a_2 = b_2 = 1$, in that case $\varphi(0, 0, 0, 0) = (f_{1,1}, f_{1,1})$.

Example 4.4. We start with an uninformative prior. Assume we have five patients in a clinical trial and we use the policy $\pi = (T_1, T_1, T_2, T_1, T_2)$. In a clinical trial a policy is often called an allocation sequence since it tells us how to allocate the trial members to the treatments. Assume further the following observations sequence $\mathcal{O} = (\mathfrak{s}, \mathfrak{s}, \mathfrak{f}, \mathfrak{f}, \mathfrak{s})$. In Table 4.1 we see the corresponding knowledge and pair of densities. Now assume that after we have observed $\mathcal{O}$

Table 4.1: Knowledge and density update

time	knowledge k_n	associated density pair	$\widehat{p_1}$	$\widehat{p_2}$
$k = 0$	$(0, 0, 0, 0)$	$(f_{1,1}, f_{1,1})$	0,5	0,5
$k = 1$	$(1, 0, 0, 0)$	$(f_{2,1}, f_{1,1})$	0,66	0,5
$k = 2$	$(2, 0, 0, 0)$	$(f_{3,1}, f_{1,1})$	0,75	0,5
$k = 3$	$(2, 0, 0, 1)$	$(f_{3,1}, f_{1,2})$	0,75	0,33
$k = 4$	$(2, 1, 0, 1)$	$(f_{3,2}, f_{1,2})$	0,6	0,33
$k = 5$	$(2, 1, 1, 1)$	$(f_{3,2}, f_{2,2})$	0,6	0,5

there is one more trial member. If we choose treatment T_1 we expect a success with probability $0, 6$. In that case we would have $k_6 = (3, 1, 1, 1)$ and the density pair $(f_{4,2}, f_{2,2})$. With probability $0, 4$ we have $k_6 = (2, 2, 1, 1)$ and $(f_{3,3}, f_{2,2})$. If we decide to use treatment T_2 then we have $k_6 = (2, 1, 2, 1)$ and $(f_{3,2}, f_{3,2})$ as well as $k_6 = (2, 1, 1, 2)$ and $(f_{3,2}, f_{2,3})$ both with equal probability $0, 5$.

4.1.1 Size and Enumeration of the Knowledge Space

First we want to calculate the size of the knowledge space $\mathscr{K}_M^4$, i.e., the number of integer valued points in the four dimensional simplex with edge length M. We can do this more generally for $\mathscr{K}_M^n$, the n dimensional simplex with edge length M,

$$\mathscr{K}_M^n := \{(m_1, \ldots, m_n) : m_i \in \mathbb{N}, \sum_{i=1}^{n} m_i \leq M\}.$$

We define $R_n(M)$ as the cardinality of $\mathscr{K}_M^n$,

$$R_n(M) := |\mathscr{K}_M^n| = \sum_{m_1=0}^{M} \sum_{m_2=0}^{M} \cdots \sum_{m_n=0}^{M} \chi_M(m_1 \ldots, m_n), \qquad (4.6)$$

where

$$\chi_M(m_1, \ldots, m_n) := \begin{cases} 1 & \text{if } \sum_{i=1}^{n} m_i \leq M \\ 0 & \text{otherwise.} \end{cases}$$

It is clear that the second equality in (4.6) is satisfied by definition of $\chi_M(m_1, \ldots, m_n)$. In the following we will find an expression for (4.6) which is easier to handle. Before we go on we need a small lemma.

Lemma 4.5. *The following equation is satisfied:*

$$\chi_M(m_1, \ldots, m_n) = \chi_{M-m_n}(m_1, \ldots, m_{n-1}).$$

Proof. Suppose

$$\chi_M(m_1, \ldots, m_n) = 1$$

$$\Leftrightarrow \sum_{i=1}^{n} m_i \leq M$$

$$\Leftrightarrow \sum_{i=1}^{n-1} m_i \leq M - m_n$$

$$\Leftrightarrow \chi_{M-m_n}(m_1, \ldots, m_{n-1}) = 1.$$

Since χ_M is either 0 or 1 we also have by contraposition

$$\chi_M(m_1,\ldots,m_n) = 0 \Leftrightarrow \chi_{M-m_n}(m_1,\ldots,m_{n-1}) = 0.$$

$\square$

Now we are able to calculate $R_n(M)$ through a recurrence relation.

Theorem 4.6. *The number of integer points in the n–dimensional simplex with edge length M, $R_n(M)$, satisfies the following recurrence:*

$$R_1(M) = M + 1 \tag{4.7a}$$

$$R_n(M) = \sum_{i=0}^{M} R_{n-1}(i). \tag{4.7b}$$

Proof. Clearly we have $R_1(M) = \sum_{m_1=0}^{M} \chi_M(m_1) = M + 1$. Now we calculate

$$R_n(M) = \sum_{m_1=0}^{M} \sum_{m_2=0}^{M} \cdots \sum_{m_n=0}^{M} \chi_M(m_1,\ldots,m_n)$$

$$= \sum_{m_n=0}^{M} \sum_{m_1=0}^{M} \cdots \sum_{m_{n-1}=0}^{M} \chi_{M-m_n}(m_1,\ldots,m_{n-1})$$

$$= \sum_{m_n=0}^{M} \sum_{m_1=0}^{M-m_n} \cdots \sum_{m_{n-1}=0}^{M-m_n} \chi_{M-m_n}(m_1,\ldots,m_{n-1})$$

$$= \sum_{m_n=0}^{M} R_{n-1}(M - m_n)$$

$$= \sum_{m_n=0}^{M} R_{n-1}(m_n).$$

In the second step we changed the summation order and used Lemma 4.5. In the last step we again changed the summation order. Replacing the summation index m_n by the index i yields the result. $\square$

Theorem 4.7. *The function*

$$R_n(M) = \binom{M+n}{M}$$

satisfies the recurrence relations (4.7) and hence equals the number of integer points in the n–dimensional simplex with edge length M.

Proof. We proof the claim by induction on n. Fix an $M \in \mathbb{N}$ and set $n = 1$. Then we have $R_1(M) = \binom{M+1}{M} = \frac{(M+1)!}{M!} = M + 1$. Now recall that

$$\binom{x}{y} = \binom{x-1}{y-1} + \binom{x-1}{y},$$

which is one of the fundamental properties for binomial coefficients. In the following calculation we iteratively apply this formula,

$$
\begin{aligned}
R_{n+1}(M) &= \binom{M+n+1}{M} \\
&= \binom{n+1}{0} + \binom{n+1}{1} + \binom{n+2}{2} + \ldots + \binom{n+M}{M} \\
&= \binom{n}{0} + \binom{n+1}{1} + \binom{n+2}{2} + \ldots + \binom{n+M}{M} \\
&= \sum_{i=0}^{M} R_n(i),
\end{aligned}
$$

hence we have proved the claim by induction.   □

In our application n is small compared to M, so we better rewrite

$$R_n(M) = \frac{1}{n!}(M+1)(M+2)\cdots(M+n).$$

This means we know now that we only can handle small clinical trails. If we have a trial size of 100 patients we already have a state space of size $R_4(100) \approx 4.6 \cdot 10^6$. The small action space and the fact that, if an action is fixed and a state is given, only two possible subsequent

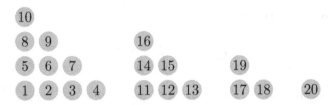

Figure 4.1: Enumeration of $\mathscr{K}_3^3$.

states have to be considered in the subsequent decision epoch still make it possible to use backward induction for $M = 100$. We only have two possible subsequent states because all the other states have probability zero.

For implementation purposes we need an enumeration of the points in $\mathscr{K}_M^n$, this is a bijective mapping $\psi_n : \mathscr{K}_M^n \to \mathbb{N}$. This is because the program represents the state space through a set $\tilde{S} = \{1, \ldots, c\}$ and we want to switch between a number in $\tilde{S}$ and a state in $\mathscr{K}_M^4$. We already know that we have to set $c := \binom{M+n}{M}$ since this number equals the size of the knowledge space. We illustrate the way of enumeration with the help of an example.

Example 4.8. We want to enumerate the points in $\mathscr{K}_3^3$. We have a look at Figure 4.1, where we can see four groups of points. The leftmost group contains all points of the form $(m_1, m_2, 0) \in \mathscr{K}_3^3$. The next group contains all points of the form $(m_1, m_2, 1) \in \mathscr{K}_3^3$, then we have the points $(m_1, m_2, 2) \in \mathscr{K}_3^3$ and finally we have all points of the form $(m_1, m_2, 3) \in \mathscr{K}_3^3$. Since $(m_1, m_2, 3) \in \mathscr{K}_3^3$ there is actually only one point of this form, namely $(0, 0, 3)$. The groups of points (from left to right) represent the intersections of $\mathscr{K}_3^3$ with the planes $z = 0$, $z = 1$, $z = 2$ and $z = 3$ in $\mathbb{R}^3$ where the point ① in the figure corresponds to $(0, 0, 0) \in \mathbb{R}^3$.

Now, for example, consider the point $\mathbf{p} = (1, 0, 2)$. Comparing to Figure 4.1 we have $\psi(\mathbf{p}) = 18$. How can we calculate this number? Well first we have to handle the z-component of the point $\mathbf{p}$. This means we have to count the number of points of the form $(m_1, m_2, 0) \in \mathscr{K}_3^3$ and $(m_1, m_2, 1) \in \mathscr{K}_3^3$, which all lie below $\mathbf{p}$. This sum equals

$R_2(3) + R_2(2)$. Then, in principal we would have to proceed with the y-coordinate analogously. Since in this example the y–component equals zero this part is dropped. Then we have to handle the x-component, i.e. we have to add $R_1(1)$ to the sum. All together we now have

$$\psi(\mathbf{p}) = R_2(3) + R_2(2) + R_1(1) = \binom{3+2}{2} + \binom{2+2}{2} + \binom{1+1}{1} = 18.$$

For an arbitrary point we have

$$\psi(m_1, m_2, m_3) = \sum_{i=0}^{m_3-1} R_2(M - i) + \sum_{i=0}^{m_2-1} R_1(M - i - m_3) + R_1(m_1).$$

This can be generalized to $\mathcal{K}_M^n$ by

$$\psi_n(m_1, \ldots, m_n) := \sum_{j=1}^{n-1} \sum_{i=0}^{m_{j+1}-1} R_j\left(M - i - \sum_{k>j+1} m_k\right) + R_1(m_1),$$

hence in the for our application important four-dimensional case:

$$\psi_4(m_1, m_2, m_3, m_4) = \sum_{i=0}^{m_4-1} R_3(M - i) + \sum_{i=0}^{m_3-1} R_2(M - i - m_4)$$

$$+ \sum_{i=0}^{m_2-1} R_1(M - i - m_4 - m_3) + R_1(m_1).$$

Evaluating these sums is straight forward but lengthy. With the help of MATHEMATICA we can calculate a closed form of ψ_4, which is a lengthy polynomial in M, m_1, m_2, m_3 and m_4:

$$\psi_4(m_1, m_2, m_3, m_4) = 1 + m_1 + \frac{3m_2}{2} - \frac{m_2^2}{2} + \frac{11m_3}{6} - m_3^2 + \frac{m_3^3}{6}$$

$$+ \frac{25m_4}{12} - \frac{35m_4^2}{24} + \frac{5m_4^3}{12} - \frac{m_4^4}{24} + m_4 M$$

$$+ 2m_3 M - \frac{m_2^2 M}{2} + \frac{35m_4 M}{12} - \frac{5m_4^2 M}{4}$$

$$+ \frac{m_4^3 M}{6} + \frac{m_3 M^2}{2} + \frac{5 m_4 M^2}{4} - \frac{m_4^2 M^2}{4}$$
$$+ \frac{m_4 M^3}{6}.$$

Nevertheless, we then have a quick way to switch from our knowledge space to the internal representation of the state space $\tilde{S}$.

Getting from a number $x \in \tilde{S}$ back to an element of the knowledge space, is done in the following way: Before we start the Backward Induction algorithm we generate a matrix $C \in \mathbb{N}^{|\mathscr{K}_M^4| \times 4}$, where $C_{x,:} := \psi^{-1}(x)$. The construction of C is described in Algorithm 7.

Algorithm 7 Number to State

Require: A number $x \in \tilde{S}$.
Ensure: $\psi^{-1}(x)$, i.e. the corresponding state $(m_1, m_2, m_3, m_4) \in \mathscr{K}_M^4$.
1: counter $= 0$
2: **for** $m_4 = 0$ **to** M **do**
3:　**for** $m_3 = 0$ **to** M **do**
4:　　**for** $m_2 = 0$ **to** M **do**
5:　　　**for** $m_1 = 0$ **to** M **do**
6:　　　　**if** $m_1 + m_2 + m_3 + m_4 \leq M$ **then**
7:　　　　　counter $\leftarrow$ counter+1
8:　　　　**end if**
9:　　　　**if** counter$= x$ **then**
10:　　　　　**return** (m_1, m_2, m_3, m_4)
11:　　　　**end if**
12:　　　**end for**
13:　　**end for**
14:　**end for**
15: **end for**

Now having access to a given row should be quick, so altogether we have a quick possibility to change between the representations of the state spaces.

Finally we have to specify the rewards. Based on the cost functions introduced in the beginning of the chapter we define the reward functions *expected successes*

$$r_i(p_1, p_2) = p_i.$$

Then we have *expected failures avoided,*

$$r_1(p_1, p_2) = \begin{cases} p_1 - p_2 & \text{if } p_1 \geq p_2 \\ 0 & \text{if } p_1 < p_2 \end{cases},$$

$$r_2(p_1, p_2) = \begin{cases} p_2 - p_1 & \text{if } p_2 \geq p_1 \\ 0 & \text{if } p_2 < p_1. \end{cases}$$

A way to use the cost models directly involving the costs of the treatments is to introduce a positive strictly decreasing function ρ and set

$$r_1(p_1, p_2) = p_1 \rho(D_1), \quad r_2(p_1, p_2) = p_2 \rho(D_2).$$

Finally, the last cost model described in the beginning of this chapter can be turned into a reward by

$$r_1(p_1, p_2) = \begin{cases} (p_1 - p_2)\rho(D_1) & \text{if } p_1 \geq p_2 \\ 0 & \text{if } p_1 < p_2 \end{cases}$$

$$r_2(p_1, p_2) = \begin{cases} (p_2 - p_1)\rho(D_2) & \text{if } p_2 \geq p_1 \\ 0 & \text{if } p_2 < p_1 \end{cases}.$$

We do not know the real success probabilities p_1 and p_2, so we take the estimators $\widehat{p_1}$ and $\widehat{p_2}$. With $\widehat{p_{1+}}$ and $\widehat{p_{2+}}$ we denote the updated estimators, that is the success probabilities in the next round. Now we can assign the rewards for the underlying Markov decision problem,

$$r(s, a = T_1) := \widehat{p_1} r_1(\widehat{p_{1+}}, \widehat{p_2}) + (1 - \widehat{p_1}) r_1(\widehat{p_{1-}}, \widehat{p_2})$$

$$r(s, a = T_2) := \widehat{p_2} r_2(\widehat{p_1}, \widehat{p_{2+}}) + (1 - \widehat{p_1}) r_1(\widehat{p_1}, \widehat{p_{2-}}),$$

where

$$\widehat{p_1} = \frac{s_1 + 1}{s_1 + f_1 + 2}, \quad \widehat{p_2} = \frac{s_2 + 1}{s_2 + f_2 + 2}$$

if we start with the non informative prior. Furthermore then

$$\widehat{p_{1+}} := \frac{s_1 + 2}{s_1 + f_1 + 3}, \quad \widehat{p_{1-}} := \frac{s_1 + 1}{s_1 + f_1 + 3}.$$

The quantities $\widehat{p_{2+}}$ and $\widehat{p_{2-}}$ are defined analogously. Now we want to define the final rewards. Assume we finished our clinical trial with M trial members. The last observation, $\mathfrak{s}$ or $\mathfrak{f}$ leads to the final state. Assume further we have an estimation μ of how many people outside the trial will receive the treatment. Then we set the final reward to $r(s) := \mu \max(r_1(\widehat{p_1}, \widehat{p_2}), r_2(\widehat{p_1}, \widehat{p_2}))$ where r_i is one of the reward models mentioned above.

Now we defined the horizon length, the state space, the transition probabilities, and some reward models and hence have a complete Markov decision problem formulation for the finite horizon case.

4.1.2 Implementation Considerations

We now want to have a quick look at the Backward Induction Algorithm 3. As already mentioned we internally represent a state by a number $s \in \tilde{S} = \{1, \ldots, \binom{M+4}{M}\}$. How we switch between a knowledge an an internal state is explained in the previous section. For a given $t \in \{M, \ldots, 1\}$ and a given $s \in \tilde{S}$ we have to solve

$$\max_{a \in A_s} r_t(s, a) + \sum_{\sigma \in S} p_t(\sigma | s, a) u_{t+1}(\sigma). \tag{4.8}$$

This can be done quickly since we only have two actions and the sum has only two non-zero terms if a is also fixed because then a state has only two possible successors. Then there is another important simplification. Consider an arbitrary history

$$(s_1, a_1, s_2, a_2, \ldots, s_M, a_M, s_{M+1}).$$

We have $a_i \in \{T_1, T_2\}$ for all $1 \le i \le M$. Now $s_1 = (0, 0, 0, 0)$, there is no other state possible since we did not start the clinical trial. In this section we denote the number of successful treatments with s_1

and S_2 in order to avoid mixing it up with the state s_1. The number of fails are denoted by F_1 and F_2. Then $s_2 = (S_1, F_1, S_2, F_2)$ with $S_1 + F_1 + S_2 + F_2 = 1$, hence after allocating the first person there are only four states possible. Histories where s_2 is a state not fulfilling this properties have probability zero and hence do not have to be considered.

If the Backward Induction is started at $t = M$, then the for-loop in Algorithm 3 has only to be executed for states fulfilling $S_1 + F_1 + S_2 + F_2 = M - 1$. Then t is set to $M - 1$ and at this stage only states with $S_1 + F_1 + S_2 + F_2 = M - 2$ have to be taken into account. Altogether this leads to an effort of $O(|\mathcal{K}_M^4|)$, hence the effort is linear in the size of the state space. Whenever there is more than one element in A_s maximizing (4.8) we assign T_1 or T_2 equiprobable.

4.1.3 Numerical Results

In this section we want to compare the Markov decision based policy π_{MD} together with the reward model *expected successes* with the outcome of an equal randomization allocation strategy π_{ER}. Now π_{MD} is a $\binom{M+4}{M} \times M$ matrix where M is the number of trial members. $(\pi_{MD})_{s,j}$ contains the information what treatment has to be used next if we are in state $s \in \tilde{S}$ at time point j. If s and j are state–time combinations which do not fit together we set $(\pi_{MD})_{i,j} = 0$. Such a blown up framework is actually not needed but it conforms with the description in the theory part.

Let $\mathcal{O}_\pi \in \{\mathfrak{s}, \mathfrak{f}\}^M$ be the vector containing the observed successes and treatment failures. Observation $(O_\pi)_i$ is a realization of a Bernoulli random variable with parameter p_1 if treatment T_1 is chosen and parameter p_2 if treatment T_2 is chosen for the i-th trial member. The treatment chosen for patient i depends on the used policy π. With

$$\text{alloc}(\mathcal{O}_\pi) \in \{T_1, T_2\}^M$$

we denote the allocation sequence given by the policy π. Now $(\text{alloc}(\mathcal{O}_\pi))_i$ contains the information which treatment person i receives. Note that $(\text{alloc}(\mathcal{O}_\pi))_i$ is dependent on the observation sequence only up to observation $i - 1$. These observations result in a knowledge (s_1, f_1, s_2, f_2). This knowledge itself corresponds to an internal state s, hence

$$(\text{alloc}(\mathcal{O}_\pi))_i := \pi_{s,i}.$$

Now we want to compare two policies. Therefore, we need a measure how "good" an allocation is. Let us define $p(T_i) := p_i$ and

$$\mathcal{F}(\widehat{p_1}, \widehat{p_2}) := \begin{cases} p_1 & \text{if } \widehat{p_1} \geq \widehat{p_2} \\ p_2 & \text{otherwise,} \end{cases}$$

where $\widehat{p_1}$ and $\widehat{p_2}$ are the estimators of the success probabilities based on the last knowledge in the trial, i.e., after the last trial member is medicated. Similar to [BE95] we choose

$$\mathcal{L}(\mathcal{O}_\pi) := (M + \mu) \max(p_1, p_2) - \left(\sum_{i=1}^{M} p((\text{alloc}(\mathcal{O}_\pi))_i) + \mu \mathcal{F}(\widehat{p_1}, \widehat{p_2}) \right).$$

We always have $\mathcal{L}(\mathcal{O}_\pi) \geq 0$. The smaller $\mathcal{L}(\mathcal{O}_\pi)$ is the better the allocation was.

Suppose we would know in advance which treatment has the higher success probability, then all patients could get the better one. This leads to a reward of $(M + \mu) \max(p_1, p_2)$ because M is the number of trial members and μ is an estimator of the people outside the trial who need to be treated. From that quantity we subtract the value of the allocation induced by the policy π. Then after the trial we need to decide how the other μ persons should be treated. This is done based on the estimated success probabilities $\widehat{p_1}$ and $\widehat{p_2}$. We simply choose the treatment which has the higher success probability. In Figure 4.2 we choose $M = 100$ and $\mu = 1000$. We furthermore fix $p_1 = 0.8$ and p_2 varies from zero to one in $0,05$ steps. For every pair of success probabilities and a given policy we simulate 100 clinical trials leading to 100 realizations of $\mathcal{L}(\mathcal{O}_\pi)$. Boxplots are a possibility to make the results visible. Every box corresponds to a pair of success probabilities and a data set of 100 loss values $\mathcal{L}(\mathcal{O}_\pi)$. On each box, the central mark is the median. The edges of the box are the 25 and 75 percentiles. The whiskers extend to the most extreme data point not considered as an outlier, i.e., observations which fulfill

$$q_1 - 1,5(q_3 - q_1) \leq \mathcal{L}(\mathcal{O}_\pi) \leq q_3 + 1,5(q_3 - q_1),$$

where q_1 and q_3 are the 25 and 75 percentiles. Outliers are not drawn.

Looking at Figure 4.2 reveals that π_{MD} works clearly better than π_{ER} if $\delta := |p_1 - p_2|$ is not too small. If however δ is small we see that the performance of both methods is equally good. If δ is less than 0.05 then it is not really possible to tell, based on a study size of $M = 100$ which treatment is superior. All in all we have a method which fulfils requirement (ii), made at the beginning of this chapter, if δ is not too small since $\mathcal{L}(\mathcal{O}_{\pi_{MD}})$ is small then. Figures A.4, A.5 and A.6 illustrate the situation for $p_1 \in \{0.2, 0.4, 0.6\}$.

The method also fulfils requirement (i) comparable to π_{ER} as we can see in Figure 4.3. If at the end of the trial $\widehat{p_1} \geq \widehat{p_2}$ and $p_1 < p_2$ or $\widehat{p_2} \geq \widehat{p_1}$ and $p_2 < p_1$ we made a *wrong decision* because the μ persons outside the trial receive the inferior treatment. The number of wrong decisions can be seen in Figure 4.3. Again we fix $p_1 = 0.8$ while p_2 varies. Furthermore Figures A.1, A.2 and A.3 illustrate the situation for $p_1 \in \{0.2, 0.4, 0.6\}$.

4.2 Infinite Horizon Markov Decision Problem Formulation

Now we want to take a look at an infinite horizon model for clinical trials. Since in this case the knowledge space

$$\mathscr{K}_\infty^4 := \{(s_1, f_1, s_2, f_2) : s_1, f_1, s_2, f_2 \in \mathbb{N}\}$$

is not finite it is impossible to use one of the presented algorithms. One possible solution is to truncate the state space and introduce

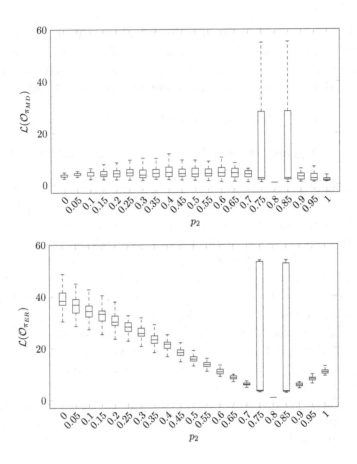

Figure 4.2: Here $p_1 = 0.8$ is fixed and p_2 varies from 0 to 1. For every p_2
value we simulate 100 clinical trials and therefore have 100 realiza-
tions of $\mathcal{L}(\mathcal{O}_{\pi_{MD}})$ and $\mathcal{L}(\mathcal{O}_{\pi_{ER}})$, respectively. Based on this data
for every p_2 a box-plot is performed.

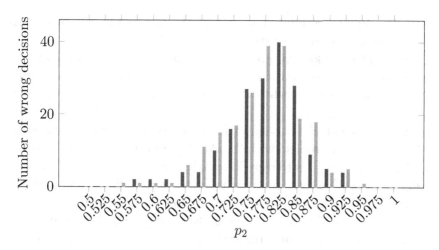

Figure 4.3: Here we see the number of wrong decisions (better treatment not detected) produced from MDP Policy and Equal Randomization out of 100 trials for fixed $p_1 = 0.8$ and p_2 varying from 0.5 up to 1 in 0.025 steps. For each value of p_2 we see one dark grey (for π_{MD}) and one light grey (for π_{ER}) bar. We see that there is essentially no difference between the number of failures produced by the two policies.

absorbing states. We again want to consider a clinical trial with M trial members. A state is an absorbing state if $s_1 + f_1 + s_1 + f_1 = M$. The transition probability for absorbing states is set to

$$p((s_1', f_1', s_2', f_2')|(s_1, f_1, s_2, f_2), a_i)$$
$$= \begin{cases} 1 & \text{if } (s_1', f_1', s_2', f_2') = (s_1, f_1, s_2, f_2) \\ 0 & \text{otherwise.} \end{cases} \quad (4.9)$$

The other transition probabilities and the actions are the same as in the finite horizon case. The rewards are in principle the same despite the fact that we have to discount them with a discount factor λ.

In an infinite horizon setting we therefore do not need an explicit estimate μ for the patients outside the trial. All patients outside the trial, in this case infinitely many, will be allocated to the treatment with the higher estimated success probability since the process is caught in an absorbing state.

Discounting is kind of problematic since then not all patients are worth the same, but: Patients treated later have therefore the advantage that they might benefit from the knowledge already learned about the two treatments.

We want to calculate an approximate solution for the infinite horizon Markov decision problem with the Policy Iteration algorithm. Looking at Algorithm 5 reveals that we have to handle two major steps in every iteration. We have to update the vector v and the decision rule d. Solving the maximization in every iteration is easy since there are only two actions, $A = \{T_1, T_2\}$. Hence updating d is easy. Furthermore, to update v we have to solve the system

$$(Id_V - \lambda P_d)v = r_d. \qquad (4.10)$$

Here P_d is the transition probability matrix induced by the decision rule d,

$$(P_d)_{ij} = p(j|i, d(i))$$

and r_d is the induced reward,

$$(r_d)_i = r(i, d(i)).$$

Note that $i, j \in \{1, \dots, \binom{M+4}{M}\}$ are natural numbers but encode, via

$$\psi^{-1} : \left\{1, \dots, \binom{M+4}{M}\right\} \to \mathscr{K}_M^4$$

a certain knowledge. Now each row of P_d contains either one or two non–zero entries. Row i contains exactly one non–zero entry if $\psi^{-1}(i)$ is an absorbing state. In the other case we have, given the decision rule d, only two possible successor states and hence only two non–zero entries. In fact we even have

Lemma 4.9. *For a given $d \in D^{DM}$ the Matrix P_d is a sparse upper triangular matrix.*

Proof. For a given knowledge $s = (s_1, f_1, s_2, f_2)$ we have four possible successor states, $(s_1 + 1, f_1, s_2, f_2)$, $(s_1, f_1 + 1, s_2, f_2)$, $(s_1, f_1, s_2 + 1, f_2)$ and $(s_1, f_1, s_2, f_2 + 1)$. Now let s' be an arbitrary successor state. Then we have that $\psi(s') > \psi(s)$ what can be seen by comparison with Algorithm 7, hence P_d is an upper triangular matrix. The rest follows from the discussion above. □

Hence, we can solve the system of linear equations by a simple bottom–up process. Beforehand we introduce the following notation: To a given decision rule d and an internal state i we denote with $i+$ and $i-$ the two possible successor internal states, $i+$ corresponds to the state if a success $\mathfrak{s}$ is observed and $i-$ corresponds to the state if a failure $\mathfrak{f}$ is observed. Let furthermore $\psi^{-1}(i) := (s_1, f_1, s_2, f_2)$. Then we define

$$p_d := \begin{cases} \frac{s_1+1}{s_1+f_1+2} & \text{if } d((s_1, f_1, s_2, f_2)) = 1 \\ \frac{s_2+1}{s_2+f_2+2} & \text{otherwise.} \end{cases}$$

Theorem 4.10. *The vector v in Algorithm 5 is updated component-wise via*

$$v_i = \frac{(r_d)_i}{1 - \lambda}$$

if $\psi^{-1}(i)$ is an absorbing state and

$$v_i = (r_d)_i + \lambda \left(p_d v_{i+} + (1 - p_d) v_{i-} \right)$$

if $\psi^{-1}(i) = (s_1, f_1, s_2, f_2)$ is a non–absorbing state.

Proof. Comparison with (4.10). □

So one iteration of the Policy Iteration algorithm needs an effort of $O(|\mathcal{K}_M^4|)$. Again the easy structure of the underlying infinite Markov decision problem makes it computationally feasible to look at cases $M \approx 100$ what yields roughly a state space size of $4, 6 \cdot 10^6$.

In Table 4.2 we find an overview for the iterates produced by Policy Iteration with $\lambda = 0.99$. We define

Table 4.2: Overview of Policy Iteration

Iterations	$\|v_{n+1} - v_n\|$	Number of indices i with $(v_{n+1})_i - (v_n)_i \geq 0.0005$
$n = 1$	96, 15	2888420
$n = 2$	8, 42	1212473
$n = 3$	2, 83	1082316
$n = 6$	0, 54	336533
$n = 9$	0, 26	33975

$$\mathcal{L}(\mathcal{O}_\pi) := \frac{\max(p_1, p_2)}{1 - \lambda} - \left(\sum_{i=1}^{M} \lambda^{i-1} p((\text{alloc}(\mathcal{O}_\pi))_i) + \frac{\lambda^M \mathcal{F}(\widehat{p_1}, \widehat{p_2})}{1 - \lambda} \right).$$

When $\mathcal{L}(\mathcal{O}_\pi)$ is small the allocation was good. In fact this formula is very similar to that in the finite horizon case, we just discounted the rewards to deal with finite sums. The rest is analogue to the finite horizon case. Note that we now do not need an explicit estimate μ for the patients outside the trial but therefore a discount factor λ. Looking at Figure 4.4 reveals that π_{MD} is again better than π_{ER}. Using the Equal Randomization strategy all data point simulated lie inside the whiskers, so there are no outliers in this case. The policy π_{MD} used for the simulation is constructed by Policy Iteration with 10 iterations.

4.3 Conclusio

We described a Markov decision problem model for clinical trials based on [BE95], [Pre09], and [Put94]. In a trial we compare two treatments T_1 and T_2, respectively. The goal is to find the superior treatment. The knowledge about the treatments is encoded in a pair—one for each treatment—of prior distributions. We restricted our attention to beta distributed priors and hence are able to encode all information in a four tuple (s_1, f_1, s_2, f_2), the numbers of successes and fails for treatments T_1 and T_2. The statistic is updated after every allocation.

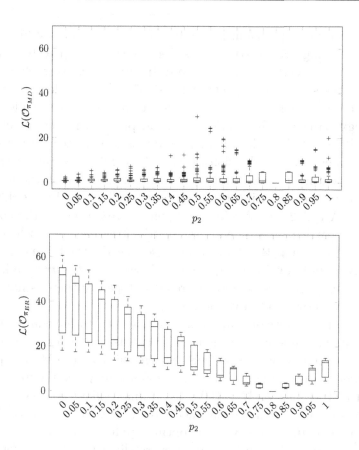

Figure 4.4: Here $p_1 = 0.8$ is fixed and p_2 varies from 0 to 1. For every p_2 value we simulate 100 clinical trials and therefore have 100 realizations of $\mathcal{L}(\mathcal{O}_{\pi_{MD}})$ and $\mathcal{L}(\mathcal{O}_{\pi_{ER}})$, respectively. Based on this data for every p_2 a box-plot is performed.

The transition probabilities are based on the expected values of the corresponding prior distributions, see 4.2 and 4.3. [Pre09] gives an overview of possible reward models. In this thesis we took a closer look at the so called "expected successes" reward model.

Based on the simulations we made we can say that the allocation strategy based on the Markov decision problem model for clinical trials is better than the Equal Randomisation strategy. First of all the number of wrong decisions, this is the number of trials where the better treatments was not detected is comparable with the number of wrong decisions using Equal Randomization, see Figure 4.3 and Figures A.1, A.2, and A.3 in the appendix. Additionally, the number of study participants receiving the inferior treatment is much smaller than in the Equal Randomization case, see Figure 4.2 and Figures A.4, A.5, and A.6 in the appendix. However if the true unknown success probabilities p_1 and p_2 are close together there is not a big difference between the two methods.

In the finite horizon framework we assumed to have an estimate μ for the patients outside the trial. If this is not the case one might prefer the infinite horizon setting. Also in this case the Markov policy is better than the Equal Randomization strategy, see Figure 4.4.

For a study size of $M = 100$, the optimal policy can be computed in a few minutes, although we then already have a state space size of roughly $4,6 \cdot 10^6$. Moreover, the optimal policy only has to be calculated once and can then be used for every clinical trial. Therefore, the solution techniques presented in Chapters 2 and 3 can be successfully applied to clinical trails.

Further developments might include improving the infinite horizon model. Moreover, the presented models might be extended to more than two treatments. In this case one needs methods for reducing the state space since then the problem becomes computationally more and more difficult.

Appendix

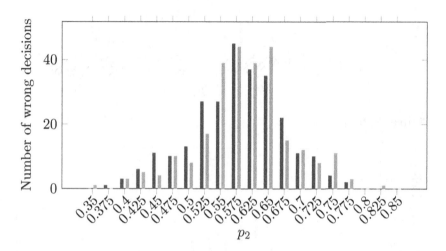

Figure A.1: p_1 is set to 0.6

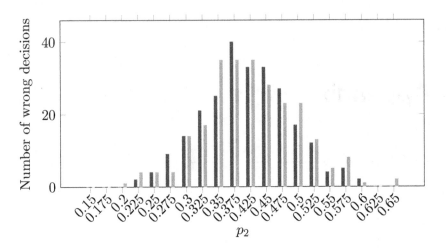

Figure A.2: p_1 is set to 0.4

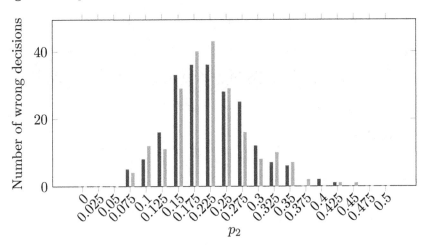

Figure A.3: p_1 is set to 0.2

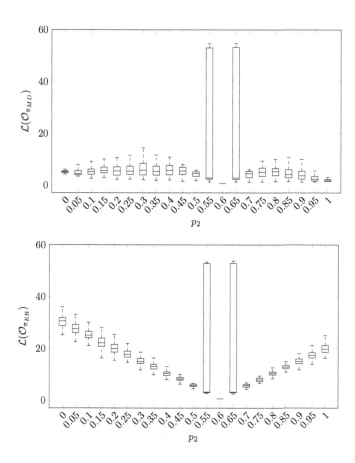

Figure A.4: Here $p_1 = 0.6$ is fixed and p_2 varies from 0 to 1.

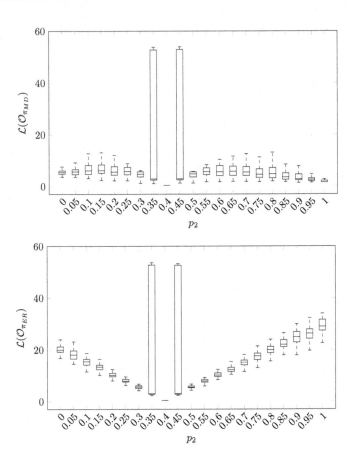

Figure A.5: Here $p_1 = 0.4$ is fixed and p_2 varies from 0 to 1.

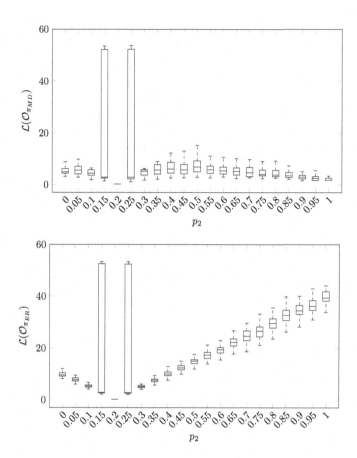

Figure A.6: Here $p_1 = 0.2$ is fixed and p_2 varies from 0 to 1.

Bibliography

[BE95]   Donald A. Berry and Stephen G. Eick. Adaptive assignment versus balanced randomization in clinical trials. *Statistics in Medicine*, 14:231–246, 1995.

[BR11]   Nicole Baeuerle and Ulrich Rieder. *Markov Decision Processes with Applications to Finance.* Universitext. Springer, 2011.

[Der70]   Cyrus Derman. *Finite State Markovian Decision Processes*, volume 67 of *Mathematics in Science and Engineering*. Academic Press, 1970.

[Geo09]   Hans-Otto Georgii. *Stochastik. Einführung in die Wahrscheinlichkeitstheorie und Statistik.* de Gruyter, 2009.

[HS91]   Janis P. Hardwick and Quentin F. Stout. Bandit strategies for ethical sequential allocation. *Computing Science and Statistics*, 23:421–424, 1991.

[Lee89]   Peter M. Lee. *Bayesian Statistics. An Introduction.* Oxford University Press, 1989.

[Pia97]   Steven Piantadosi. *Clinical Trials. A Methodologic Perspective.* Wiley Interscience, 1997.

[Pre09]   William H. Press. Bandit solutions provide unified ethical models for randomized clinical trials and comparative effectiveness research. *PNAS*, 106(52), December 2009.

[Put94]   Martin L. Puterman. *Markov Decision Processes. Discrete Stochastic Dynamic Programming.* Wiley Interscience, 1994.

[Ros99]   William F. Rosenberger. Randomized play-the-winner clin-
           ical trials: Review and recommendations. *Controlled Clin-
           ical Trials*, 20:328–342, 1999.

[RSI⁺01]   William F. Rosenberger, Nigel Stallard, Anastasia Ivanova,
           Cherice N. Harper, and Michelle L. Ricks. Optimal adaptive
           designs for binary response trials. *Biometrics*, 57:909–913,
           2001.

[SAM04]   David J. Spiegelhalter, Keith R. Abrams, and Jonathan P.
           Myles. *Bayesian Approaches to Clinical Trials and Health-
           Care Evaluation*. Statistics in Practice. John Wiley and
           Sons, 2004.

[Sen07]   Stephen Senn. *Statistical Issues in Drug Development*. Stat-
           istics in Practice. John Wiley and Sons, 2007.

[SR02]   Nigel Stallard and William F. Rosenberger. Exact group-
           sequential designs for clinical trials with randomized play-
           the-winner allocation. *Statistics in Medicine*, 21:467–480,
           2002.

[SR13]   Oleksandr Sverdlov and William F. Rosenberger. On recent
           advances in optimal allocation designs in clinical trials.
           Journal of Statistical Theory and Practice, 7:753–773, 2013.

[Whi78]   D. J. White. *Finite Dynamic Programming. An approach to
           finite Markov decision processes*. Wiley Interscience, 1978.

[Whi93]   D. J. White. *Markov Decision Processes*. John Wiley and
           Sons, 1993.

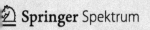